Alien Secrets and the Path to Ascension

Scott Campbell

SECTION: FRONTMATTER

ALIEN SECRETS AND THE PATH TO ASCENSION: NOTE FROM AUTHOR SCOTT CAMPBELL

Modern science, with its innovative technology, coupled with a close look at the Genome Project, informs us that our DNA proves to be genetically engineered by an advanced alien civilization. DNA applies mathematical principles to optimize performance and maintain its basic structure and function for billions of years. We have now unlocked its virtual codes. All life on Earth is an alien hybrid. The evidence remained undiscovered for over three and one-half billion years.

Alien microbes, especially the Giant Viruses, played a huge role in our evolution.

Whether intentional or not, the aliens have already presented us with clues on how to solve two of the most challenging problems we face. Both involve the mastery of quantum mechanics. Einstein could not figure out, after thirty years of attempts, nor many of the top physicists for the last one-hundred years, how to come up with a workable theory for quantum gravity that meshes with the theory of General Relativity, the so-called God Equation." [1]

The aliens did because they applied that knowledge to propel their ships by curving space-time around them. The aliens influence our consciousness and objects from a distance. Hints about their levels of prowess also came with the little, gray ET clones found at the Roswell crash. The "biological entities," rather than mere pilots, were part of the ship, a single unit of impressive technology.

The key to fully understanding ourselves lies in understanding how nonlocal quantum effects in the neuron microtubes underlie consciousness. [2] The aliens seemed to have mastered this as well. Understanding quantum mechanics and alien technology will help us "ascend" to levels comparable to the aliens at an accelerated pace. Perfect health, higher intelligence and empathy, and an extra helping of longevity will all be within reach of science and medicine. Free—or at least very--cheap and high-powered energy, and also extremely disruptive to a fossil fuel planet, is a given barring censorship.

PREFACE

We thank the US Navy for releasing authentic data on several observed spacecraft--now termed UAP for "Unidentified Aerial Phenomena." The Navy captured the data, which shows how the UAP executed their astounding speeds and navigation, leading to a general theory of how UAP moved. The videos come with chain-of-custody authentication, meaning they show originality free from any form of tampering.

Einstein would stand proud because the aliens use a principle based on geodesics that he described at length in his published papers on General Relativity.

Aliens have dominated life on Earth since day one:

- *Life on Earth arose from genetically engineered coding in alien microbes.*

- *In addition to their spacecraft's astounding and mind-boggling motion, aliens have other surprising skillsets as the ultimate scientists. Aliens are experts at cloning. The Roswell pilots were genetically engineered clones for inhabiting and piloting spacecraft. They can erase selected memories in Humans, install implants, communicate telepathically, shut down or manipulate electronic systems like missiles, temporarily paralyze, beam people or objects into their ships, and control space-time through the mastery of physics at the quantum level. They likely apply quantum AI and quantum computers to achieve some of their technological prowess.*

- *Based on ancient history and myths, aliens influenced our first civilizations and ours now, as well as everything in between. They were perceived as gods, which led to the Abrahamic religions that replaced ruling councils of aliens with concepts of a monotheistic god.*

- *Some aliens have expanding hybridization programs based on mass abductions.*
- *Some aliens have shown hostility regarding our proliferation of nuclear weapons and space exploration, particularly the moon.*
- *The technology extracted from the crash at Roswell and other crashes was behind our revolution centered on Silicon Valley.*
- *Alien mastery of quantum consciousness and cloning could lead to various new forms of being or "afterlives."*
- *Some aliens plot to infiltrate our populations with disguised superiority.*

Takeaway: Studying alien technology, consciousness, and skillsets will lead modern Humans to a similar mastery and will not only help guarantee our survival as a species but guide us into multiple levels of ascension.

For national security, maintaining the status quo, and expanding global power and profits for large corporations, the United States government decided long ago with the first retrieved alien craft not to share what they knew.

After seeing the Roswell crash reported on the front page of a local newspaper from New Mexico, USSR's Stalin sent an army of agents to find to spy on us. [8] They would work to assist communists here in creating our CIA and infiltrating it. [9] To help shut out Stalin and keep the treasure for themselves, our government's official position on UFOs and aliens soon became one of denial and disinformation:

"What UFOs?"

"What aliens?"

"The crash at Roswell was not a UFO, but a balloon, and the bodies retrieved were not four-foot-tall, gray alien clones with oversized heads and eyes, and skinny appendages, but test dummies, chimpanzees, or Japanese spies. Case closed."

The Cold War and the technological prowess and huge profits to be reaped from Roswell overshadowed officially informing the public about the greatest discovery in Earth's history. Instead, we settled for B-rate movies about aliens and UFOs. — until, in 1971, twenty-four years after Roswell, we finally welcomed the classic Close Encounters of the Third Kind, based on real-life encounters. Ufologist and astronomer J. Allen Hynek consulted for the film. [10 11]

We had entered a cat and mouse game with the Soviets since 1947, with the secrets of aliens and their craft and powers as the mouse to be captured, that never ended. In this book and The Day After Roswell, Heroic Lieutenant Colonel Corso details how his every move to help the timely dissemination of certain alien artifacts and technologies to American industry found itself shadowed by the CIA and KGB. The CIA—and the Air Force—just happened to be formed within three months after the Roswell crash in early July in 1947. [12]

Corso once walked into the office of the Soviet communist sympathizer in our military, whom he was sure was behind ordering the spying, and laid down a .45 pistol on his desk. Corso intimated that the next person who followed him might have his eyes shot out.

In addition to a cold war with the Soviets over UFOs and aliens, our military-experienced an internal cold war between its departments concerning who would get the most goodies- and hence benefits-from Roswell.

The secrets of Roswell were about technology, weapons, global power, and trillions in future profits. The American people were lied to and continuously disinformed, often with most insulting explanations—'It was a weather balloon, or swamp gas, or reflections of light off of geese undersides."—were not to ask questions but only to acknowledge the derived products like lasers, integrated circuits, transistors, fiber optics, particle beam weapons, and small and powerful personal computers and cell phones. We eagerly chose to participate in new, robust, and earth-shaking tech industries which arose from Roswell artifacts.

Washington, DC, found \ the setting for a flyby by UFOs in 1952. Photographs from multiple perspectives, hundreds of witnesses, and numerous radar reports confirmed it, but the government dismissed it as "reflections."

The purpose of Project Blue Book was to reassure the public that they had a system for reporting what they witnessed and to distract them from the Roswell crash.

Grudge and Sign were classified projects with a greater level of secrecy to enable the military to evaluate sightings and encounter reports that could not be explained away as balloons, geese, or the planet Venus. Blue Fly and Twinkle, along with a slew of other camouflage programs such as Horizon, HARP, Rainbow, and even the Space Defense Initiative, all had to do with extraterrestrial technology. However, few stood aware of it.

Smart people saw the Roswell crash and the increasing number of sightings and legitimate abduction cases through the government's coverup efforts. They realized that we were not alone --if they didn't know already. Alien craft hovered above New Mexico area nuclear missile production and testing sites, power plants, and naval ships, like insects to light at night.

GOOD NEWS AND BAD NEWS

Let's get through the bad news first. The government intentionally misled the public about aliens and UFOs for at least nine decades and likely continues to censor key information. For those of us who were not smart and deductive, the answer to one of the most important questions for Earthlings about possible galactic company lay buried in government bureaucracy. It turns out that the other question most frequently asked by Earthlings is, "Where did we come from? The truth may, again, hide among alien secrets. Science is telling us that somebody or something genetically engineered our DNA over 3.5 billion years ago. In Section: DNA, we will explore the mind-boggling and exciting evidence in detail.

The good news: The United States successfully kept most of the alien technology and insight from Roswell and other crashes out of the hands of the communist Soviets and Germans.

The Pentagon UFO films are selected visual recordings of cockpit instrumentation displays from US Navy fighter planes deployed aboard aircraft carriers USS Nimitz and USS Theodore Roosevelt in 2004, 2014, and 2015 and additional material captured by other Navy personnel in 2019. [13]

Since 2017, the four monochrome films scooped up considerable attention, as they officially captured UFOs on vetted media. In 2020, the Pentagon addressed and publicly published the first three films, and, in 2021, the Pentagon verified the origin of the leaked 2019 footage. [14] Commander David Fravor examined radar signals of a potential target off the coast of southern California on November 14, 2004. [15] The USS Princeton (CG-59), which was part of the strike group, had

GENESIS: ANOMALIES, ANOMALIES, ANOMALIES

Genesis 1.1

The Hebrew and Greek of the Septuagint versions of the Bible refer to God as "Elohim," a plural noun, implying more than a single God.

Genesis 1:26

"Let us make?" as in "Let us make humans look like us-in the images and likenesses of ourselves?"

The New Jerusalem Bible suggests that "image" implies a physical resemblance, like the son of Adam and Ada. "Image" reflects a term that is "concrete."

Again, we have a plurality, "us," and not a single God.

Genesis 2:10-14

Why does the described garden of Eden happen to reflect high-grade deposits like gold, onyx, and resin? Who would make effective use of those materials?

Genesis 2:17

The Tree of Knowledge is forbidden.

If God wanted humans to love and have free choice, why deny them the knowledge of evil versus good, which seems a prerequisite for choosing mates and a destiny? Is not knowledge useful for survival? Did God rig life for failure? Picking from the tree of knowledge is punishable by death. Does this sound like a loving and just God?

Genesis 2:21

God names the local fauna, but which offered up no suitable helpers. Seeking a capable helper, God put the first "man" into a deep sleep, extracted a rib, and used the flesh to patch him back up. From the rib, he fashioned a helper, "woman."

Is the all-knowing God short-sighted?. God, the ultimate source of all wisdom, forgot to make women. Much to modern woman's chagrin, women earned the subservient role of "helpers" only after animals were deemed incompetent.

Genesis 3:1 Now the Snake...

But the first man has arms and legs, has a brain that can handle language. "Man" does not include talking snakes. But an intelligent and talkative snake shows up out of nowhere. The Snake has a one-track mind to convince Eve to partake from the Tree of Knowledge, and, of course, she does to become conscious.

(Author's note: Genesis 1:24-27 teaches that man came into being after the animals. Genesis 2:18-19 days before.)

Genesis 3:16 Women will give birth, but with lots of pain.

Is this a mere consequence or punishment? There is to be no childbirth without it. All of that pain will come after consciousness, sexuality, and self-awareness.

Genesis 3:22 And Yahweh Elohim pointed out that the human became similar to one of the—plural. The word "God," again, represents a group.

Genesis 4:14 Cain is driven from the Garden of Eden: He will be a wanderer, restless, on the Earth, and whoever finds him will kill him. If Cain wanders beyond the garden, he is a goner. But who will kill him? It can't be Adam and Eve's children.

Genesis 6-9: The animals pair up.

are weather balloons, so the government had revealed the seriousness of the subjects of their investigations.

Politico soon claimed that Elizondo, keeping his non-disclosure agreement in mind, nevertheless admitted that Navy pilots witnessed dozens of unexplained sightings. The "threats" showed astounding maneuvers and speeds way beyond any known aircraft. Ships at sea, nuclear facilities, and power plants were among common sighting locations. Although such sightings have been reported, including reports by military personnel worldwide since WWII, Elizondo quipped that they had never witnessed anything quite like it.

Bigelow Aerospace earned a contract to examine materials that came from UFOs. Some of the materials were rumored to have few similarities to any on Earth. One is purported to be a "memory metal" capable of remembering its original shape after being folded without being heated like "Nitinol," a modern alloy of titanium and nickel. A remarkably similar material occurred at Roswell in 1947. Alien spacecraft may change size and shape during changes in direction or speed, and memory metals become quite useful for restoring original dimensions.

Of all of the pages in the 154-page report, these "materials" found and under examination, if real, are sufficient evidence that aliens visited Earth, especially if they are composed of elements not found on our Periodic Table. Some appear to be forms of isotopes not found on Earth. One investigator for Bigelow Aerospace stated publicly in a video that he would like to start utilizing such materials in various applications.

According to Popular Mechanics, Eric Davis, an astrophysicist, consulted with the original UFO program from the Pentagon and informed a Department of Defense agency regarding off-world vehicles and materials that Earthly scientists could not fabricate. [17]

Some people claim they suffered health problems after encountering a UFO, and the government admitted to studying them but fell short of revealing their findings.

What is still blasted all over the internet is the footage of the November 2004 encounter of UFOs by the USS Princeton within one hundred miles of San Diego Pilot David Fravor, the most quoted witness in the media, concluded that the craft was not of this world.

To have the Navy admit that they saw this, much less show footage of it, is indeed promising progress. Congress did pass a bill saying that the Pentagon was to come clean about what it knows about UFOs. The government said more information is to be disseminated by June 25, 2021. Other videos show UFOs submerging themselves in the ocean. The government said it had no proof that the UFOs were alien but couldn't rule it out. An explicit order tucked away in the Senate Intelligence Committee's Intelligence Authorization Act for Fiscal Year 2021 found itself part of the $2.3 trillion COVID relief Bill passed in December of 2020. It states that the Pentagon must disclose what it has found regarding UAP so far by the end of June 2021. It didn't happen as the public wanted.

SECTION: AGENDAS AND IDENTITIES

AGENDAS

Given the evidence that aliens increase their presence on Earth, we must address what they want and their identity. These questions become two of the most important to ponder because aliens, like it or not, either by their actions or by our reactions to them, established themselves as a major evolutionary force on ancient, pre-modern, and modern Earth.

Extensive research on the most realistic alien encounters to support current theories is now available. Accounts of UFOs that include physical or electronic evidence, multiple witnesses, and government records that have been declassified offer the best raw data to validate spacecraft's existence. Cases of alien contact, including abductions, seemed to be intentionally muddled by the aliens from the start. The abductees usually experience trauma, their bodies often become paralyzed, and their memories are selectively erased to the point where hypnotic regression becomes the best and only way to get at what happened. "Wholly overmatched" is how some describe the human-alien interaction in which the aliens display capabilities for cognitive management, the powers of their amazing technology, and vastly superior intellect. Controlling our emotions seems straightforward for them. They reign as the ultimate experts at stealth. Selectively impairing or erasing abductees' memories does not facilitate easy and fluid recall of the event. Why would the aliens want to hide what they are doing to humans?

Abductees who would prefer to forget the incident altogether or label it as a dream, or a result of their imagination, do not help find the truth through investigation either.

Although the total amount of information doctors and therapists glean from encounters shows anticipated variation, as it should. An undeniably broad similarity and consistency in abductions revealed itself. We net out and sift through vital details about the nature of aliens, their actions concerning humans, their abilities, and their craft—and their agendas.

Reported cases number well into the thousands, and millions of people may have been abducted so far.

Refuting the presence of alien craft on Earth is much more difficult now, with released military videos of UFOs demonstrating unheard of maneuvers that scream "aliens!"

Either the craft represents aliens or alien technology that has been reversed engineered. Either way, it is evidence of an alien presence. (The aliens demonstrate they understand and apply a complete quantum theory of the universe which so far has stumped humans, rendering preposterous the notion that the craft represents sources of black-box operations or Chinese or Russian technology.)

We can focus on human-alien interaction cases that clothed themselves with validity and rich amounts of information to formulate agendas and describe alien populations.

The presence and influence of aliens cannot be fully understood without considering our current trajectory as a species.

Understanding the aliens means understanding ourselves. Mainstream media's altitude regarding UFOs progressed since the release of military tracking starting in 2017. Talking and writing about UFOs gained a new air of respectability.

WERE WE ENHANCED?

Truth to any ancient alien hypothesis would have a great bearing on agendas. See Section: DNA which presents biological, evolutionary, and mathematical evidence that life started on Earth from alien genetically engineered DNA.)

Dr. Colleen Clements, a former professor at the Rochester Institute of Technology and the University of Rochester, authored two books regarding ET ancient visitation and that humans are hybrids with aliens. She thinks the main enhancement in our evolution occurred some 40,000 years ago. She cites both mythological and genetic evidence. Although disputed by other scientists, Clements argued that our microcephalin gene is crucial to larger brain size than the rest of our primate friends. The allele originated during a likely span of 14,000 to 60,000 years ago in central Eurasia. The Denisovans, also from central Asia, serve as another candidate. Denisovans mated with Homo sapiens but did not go overboard. However, the D allele was not found in Denisovans either.

AS RECORDED THROUGH ARTWORK, the D allele may have triggered an explosion in human invention and creativity by enhancing brain development.

Not everyone today has the D allele. As you go farther from north-central Asia, you see less of it.

The first sky god myths or "skydiver" myths appeared around the same time as the D allele and in the same area, north-central Asia.

Theory: the sky gods were aliens and bestowed higher intelligence upon the early humans at that time.

Similar myths spread from that same location around the world. Old texts speak of an extraplanetary species called the

Bene Ha-Elohim, Anunnaki, Watcher, or one of many other names.

MONITORED OR CONTROLLED?

Aliens may have handed one-way tickets to early humans to join them in off-planet excursions and civilizations. This would follow logically after modifications.

Records of ancient UFO sightings prove to be rare, and few depict details of the appearance of the technology. Instead of metallic saucers or cigar-shaped craft, we get accounts of misbehaving fireballs. "Fireballs" serve as an apt description for UAP because they emit light, sometimes bright, as part of the distortion fields that embed them from their modulation of space-time near their craft. Accelerating UAP prove to be lit up shape-shifters rather than confined to rigid "cigars" or "saucers."

A long-term alien presence is more likely than not, even if it remained covert most of the time—until Sumeria reported them as a ruling council.

Australia's aboriginals depicted tall whitish skydiver gods, wearing robes in cave paintings rather than little gray aliens with big heads and eyes. Aboriginals never wore robes and were never White. Diminutive grays seem absent in history until WWII and after becoming one of the most common encounters.

CONTACT NOT LOGICAL

Ancient people lacked the intellect to comprehend the true nature of alien visitation fully. There would be no great benefit of aliens making attempts at extensive communication or education and seeing little for them to gain in trying to do so. Modern times are a different story.

HUMANITY'S REVOLUTION

Hunting and gathering were the only two occupational choices for us for 290,000 years as Homo sapiens, what Richard Dolan labels as Stage One existence in The Alien Agendas. Ten thousand years ago, we started to farm and domesticate plants and animals. Stage Two arose from an agricultural revolution that allowed larger populations to take root alongside their crops and stay in one area.

We netted larger populations to underlay towns and cities, new organizational methods, and innovative tools to benefit humankind:

- Irrigation
- Writing
- Use of the wheel
- Metallurgy
- Stability
- Division of labor
- Domestication of plants and animals
- A flourishing of the arts
- The unit worker—who could thrive in a variety of cultures in a variety of locations.

Stage Three

- Based upon the 17 th-century scientific revolution spearheaded by Leibnitz, Newton, and Descartes and embedded in the hypothesis/test/theory protocol.
- Industrialization, the application of mechanization to create services and goods rather than raw muscle.
- Although some of it may have been derived from alien technology, for example, from Roswell, the scientific

method sped up a series of key inventions, all the way up to computers, Wi-Fi, satellites, and nuclear power, and nuclear weapons.

- Increased UFO sightings starting in the 1750s, coinciding with the industrial revolution, accelerating in the mid-1940s after nuclear explosions, and continuing through today, facilitated first by radar, then by infrared cameras, and now by 3-D radar and sonar.

Fourth Stage

This is Dolan's Transhumanist Stage to be described in detail later.

ONE FINAL QUESTION ABOUT OUR POTENTIAL ANCIENT ORIGINS

Aliens may have tweaked our DNA and fashioned a non-biological command structure, an artificial intelligence structure akin to a living machine, to account for the consistent influence of the human populations. Such a "non-biological" entity would follow directives:

Such a system would:

- Prevent changes that would put the AI at risk.
- Control any threats it could not prevent.
- Emphasize stability rather than rapid advancement

ALIEN HUMANOIDS

SOURCES?

From perusing unclassified government documents worldwide, we can conclude that advanced forms of science and technology exist that are not supposed to exist. There is no evidence of human development of super aircraft capable of UFO-type maneuvers. Little is mentioned in the documents about the nature of craft occupants.

Witness testimony about UFO crashes and retrieval is such a large body of evidence with such quality and quantity that it is hard to think that aliens do not exist. Although many people report encounters, or CE, "Contact Experiences," the evidence varies widely in reliability, validity, and quality.

THE FREE STUDY

Edgar Mitchell, the well-known and outspoken astronaut who explicitly stated that he believes aliens exist, is the impetus behind his Foundation for Research into Extraterrestrial and Extraordinary Experiences or "FREE." Thousands of experiencers reported contributing to the study, performed through unfiltered online questionnaires. Despite the possibilities of some bogus reports, the overall results hold solid value. [18]

Edgar Mitchell Foundation Study: : 1) Small gray (67%), 2) Aliens or non-human intelligence (NHI) (64%), 3) Energy Beings
(55%), 4) Spirit forms (50%), 4) Tall gray (47%), and
5) Hybrids (40%). [19] [20]

According to the FREE study, only fifteen percent of the contact experience respondents, including all types of experiences beyond abductions, had an overall negative emotional experience. Twenty-three percent registered their experience as neutral, and sixty-two percent said it was positive. For Out the body experiences, the positive evaluation went up to seventy-one percent.

Many aliens were perceived to have humanoid forms. He conjectures that our physical forms are logical and evolutionarily advantageous, and where conditions are similar in the universe, one might think that human-looking beings might evolve by chance.

Half of the sightings were of aliens that looked very human.

How would that happen?

Perhaps Earthlings were monitored, enhanced, and transported to other planets to form hybrids. This could have happened from 60,000 years ago to a few thousand years ago or even more recently.

We can assume such hybrids would boast telepathy, enhanced control over consciousness, better mental abilities, enhanced perception, better health overall, and longer life spans.

According to Army Sergeant Clifford Stone at the 2001 National Press Club meeting on UFOs in Washington, DC, there were at least fifty-seven alien humanoid species living on Earth. Stone's assignment in the Army had been to help recover crashed UFOs.[i]

AVATARS AND HUBRIDS

Some humans could be infused with alien consciousness—making them undetectable Avatars.

A well-known abduction researcher, Dr. David Jacobs, proposed a hybrid-hubrid theory. Insectalin aliens led gray alien teams to create an alien-human hybrid identical to a normal human in looks but boasted keen telepathy and mind control abilities. These entities integrated into society in the last century or so. He claims they plot nothing less than a systematic infiltration of humanoids in the Great Change and have experience doing so on other planets. A powerful and open presence eventually becomes integral to the plan.

GRAYS

Large heads, skinny bodies, appendages, oversized eyes, slightly slanted upward toward the exterior, wrap-around to infringe upon the side of the face, all characterize the grays. The tiny grays earned their spot, as most people think of when "aliens" come up as a topic. Ears, nose, and mouth prove to be extremely small or nonexistent.

When you assemble the tall and short grass, they slightly outnumber human-looking aliens in reported sightings. Both groups together, as we have discussed, make up the vast majority of alien encounters.

To no one's surprise, gray appearances vary by the shape and size of their heads and the exact color of their skin, from dark grey to white, to brown, to even beige with a pinkish overtone. Some have three fingers, some four.

Even the eyes can vary in shape and color, but most are pitch black. The eyes of the tall grays find themselves described by humans as arresting or hypnotic, representing potent powers of telepathy and emotional projection and reading. The talls prove to be extremely thin. No human ever witnessed grays eating. Grays are believed to absorb nutrients through their skin and their suits.

Grays earned a reputation as speed demons but did not run as we do. They glide. Leaping over fences or other barriers is nothing for them. Physical prowess through technology is suspected, and so is telepathy and mind control. They are said to control emotions and behavior in humans, helpful in the abduction projects. Selectively erasing memory masks their agenda. This would be accomplished most efficiently with applied technology based on quantum and electromagnetic elements of consciousness.

Little grays seem constrained to neutrality in gender. These little worker-beings, or gray minions, stand accused of being genetically engineered. Tall grays show male or female gender, not via genitalia, but through projection. Emotionally empathetic and caring are female stereotypes. Males seem more aloof, observant, serious, and scientific. No human ever claimed to see grays mating.

Minion grays act as pilots, groups of abductors, and micro-managing humans. They are employed by cells that seem better at communication with humans and perform more serious surgery.

Clothing? Skin-tight jumpsuits which appear spun-on rather than put-on is one choice. Nothing at all serves as the alternative.

The crafted interior often fits consensus descriptions with mysteriously diffused lighting and metallic, smooth, and curved walls. A control room is common, with viewing screens, panels, and benches attached to walls. Walls may become transparent in the blink of an eye, and equipment can pop out from anywhere. The overall size of the interior seems much larger than the perceived exterior would allow.

One abductee learned that the craft is powered by contracting and expanding space-time, moving it rather than the craft.

(Physics analysis of UAP videos and radar data of military sightings confirms this hypothesis, allowing the craft to move along geodesics free from the effects of acceleration.) Grays seem to bond themselves to the craft for its control, often with headsets full of electronics.

Some abductees claim to have been taken to large caverns in the Earth for their procedures after traversing through solid rock or ground.

Grays are known as head illusionists, sometimes appearing in clothes to the abductees. They can control gravity, energy, matter, space-time in their immediate vicinity, and let's not forget cloaking themselves as invisible.

In 1947, after the Roswell crash, Lieutenant Colonel Corso will claim in Section: Roswell that we began a form of war with Grays that led to the Strategic (Star Wars) Defense Initiative. Controversial Phil Schneider, who worked on building underground bases for the US government in Dulce, contended that they were at war with underground Grays since 1979. He participated, losing two digits off one hand, all of his toenails, and taking a wound to his chest after being hit by an electromagnetic beam weapon produced by a tall gray adorned with the weapon fastened to his chest. Schneider's assessment is dire, saying that he became informed of eleven alien races on Earth, two of which were benevolent, but one of which left town. The remaining nine were hostile. He also claimed that the people behind the New World Order proved to be aligned with some of the belligerent aliens who sought control of the planet and the death of a majority of its population by 2029. He likened the "experiments" on Humans to those of Nazi death camps. [21]

Schneider also shared that a number of materials derived from alien technology became of common use in the military for planes and submarines construction.

WHY ABDUCTIONS?

Grays, not human-looking aliens, stand suspected of abducting most humans. Given that grays often treat people well and even heal their ailments, many people hold no grudges.

He asserts that human-alien hybrids serve as the grays main goal. They became experts in human reproduction and often employed intrusive methods to obtain ova or extract sperm.

The night is the time for abduction, and when people are alone or surrounded by only a few people who can be rendered unconscious.

Tiny implants prove to be a common claim, and the process can be painful. The implants are said to emit a particular frequency of radio waves. They could provide tracking data, but there could be other applications, like recording emotions and perceptions. Some believe the grays are duplicating their brains into ours.

The more powerful beings, the insectalins or tall grays, will perform some mind penetration at least once during the abduction process to selectively erase the memory and perhaps hypnotize to control behavior later.

Tanks or aquariums that house fetuses in suspension, or even children or adult hybrids, is not an uncommon perception.

Forced mother-child interactions are a common experience, as if the hybrids need that bond to mature properly.

Abductions can be terrifying despite the aliens' efforts to make them less so-- if they feel so inclined.

Many abduction experiences are terrifying. One man claims he received a probe up his nose into his brain and experienced excruciating pain during the procedure and weeks afterward.

BRAIN ANOMALIES IN CONTACT EXPERIENCERS

When examining the MRIs of people who claimed to be close to aliens and their devices at least on one occasion, Stanford scientist Gary Nolan observed an overflow of neurons amidst the basal ganglia. Nolan noticed these experiencers weren't harmed but rather had an overflow of neurons linking the caudate nucleus's head to the putamen. The caudate nucleus controls motor planning, learning, and execution; the putamen influences motor planning, learning, and execution. If you looked at 100 ordinary persons, you wouldn't notice this type of density. These folks, on the other hand, were in possession of it. The question is whether or whether it was caused by coming into contact with something presumed alien.

Everyone has this connection region in general but suppose the typical person's density level is 1x. The majority of the subjects in the research had typical density in this region of 5x to 10x and up to 15x. We're presuming that density here refers to some form of neural activity. These people were "gifted" in terms of high and efficient levels of basal ganglia functioning.

Nolan doesn't think their UAP experiences caused their unusual brain structure but we might speculate that it drew alien attention. Abductees often report multiple events over a lifetime. What if the brain enhancement occurred gradually over the years after visits to alien spacecraft as a child? It seems remarkable that many of the UAP experiencers he studied had the exact same type of super-charged basal ganglia.[1]

---

[1] https://www.vice.com/en/article/n7nzkq/stanford-professor-garry-nolan-analyzing-anomalous-materials-from-ufo-

crashes?fbclid=IwAR1Dkgo4oXt_nmP0i1ElnEjFV2tD1724Vpai8k-vqLOlWaeCZo1YfyXgyEI

TELEPATHY AND LOW-FREQUENCY BRAIN WAVES

Our brain waves range from 5 to 30 Hertz. Higher levels coincide with higher levels of consciousness. Lower levels coincide with meditative states and states of rest. The weakest frequencies facilitated telepathy. With their applied tech, the aliens could utilize neural inks.

THE GRAYS AND THEIR AGENDA

Consensus: the hybridization projects seek to hatch human-looking hybrids bent on infiltrating leadership in societies and taking over planet Earth.

Of course, the aliens might tell their abductees a different story. They are here to help, stop us from destroying ourselves, cease environmental devastation and nuclear war, and elevate consciousness and spirituality. (One cannot help but think of Communist methods for comparison—promise a utopia with shared wealth but deliver a totalitarian state full of destruction and wealth extraction.)

No matter the ultimate purpose of their agenda, most Ufology researchers agree that hybridization is taking place on a large scale.

Theory: Grays and perhaps some human-alien species are already here among us and executing their plans.

MANTIDS OR INSECTALINS

Although small as a percentage of total aliens perceived by abductees, insectalins or mantids may be the most influential and powerful. Unlike grays, mantids offer challenges to perception. Only extended meditation or out-of-body experiences lead to an encounter, by definition dream-like.

No matter how rarely they are encountered, the experience is consistent across observers. Mantids range from a height of six feet to eight feet or even taller. Black eyes, or dark eyes of other colors, perch upon a triangular head. Mantids evolved or were engineered to have more joints than we do and thin and long bodies. A darkish brown is the most common color, but some are black or green. If they wear clothes, it is robes with hoods or collars.

It's rare to encounter a commanding mantid without those they command, the grays, around. Might the taller Grays also be genetically engineered so as to be easily tracked and controlled? Like the tall grays, mantids excel at telepathy and project gender. Mantids place themselves higher on the alien totem pole than tall grays.

Mantids telepathically engage human abductees by demonstrating impressive abilities to ease fears and emotional distress and form striking images in human minds. Projections may come across as extremely negative, such as planetary destruction. Tall grays do the same, suggesting they share a common culture and agenda. Clicking sounds allow for mantid-to-mantid intense communication.

Their ability to project images is so good that they can appear human. They seem more adept at this than any gray. Mantids have a history of occasionally engaging people without a Gray entourage and when people are fully awake.

Mantids conjure up impressions as kind and gentle spiritualists bent on helping and healing humanity. Some believe they direct the human hybridization projects as well as the manufacturing of grays. Mantids want mantid-like empowered human-alien hybrids to run Earth's leadership like they run the grays.

GRAY AND MANTID TIMELINE

No evidence of grays and mantids tag-teams appears in the literature until the late 1940s and 1950s. The recent arrival to Earth by the mantids, grays, and reptilians are synchronized with our nuclear weapons discharges and technological and scientific rapid advancement.

Colonel Corso, in Section: Roswell, attributes these advancements to the reverse-engineering of Roswell crash artifacts. Nuclear discharges drew lots of alien craft for surveillance which led to one or more crashes in a lightning storm. According to Colonel Philip Corso, the Roswell craft held five small grays that had been genetically engineered. The UFOs were labeled a hostile threat, and the military had orders to shoot them all down. A form of war began between the grays and humans. Could the mantid-gray hybridization projects result from that war and a blatant attempt to take over planet Earth from within civilization?)

EBENS

Extraterrestrial Biological Entity or "Ebens" present themselves as short grays, about forty-eight inches tall, in the 1950s and 1960s but without the typical huge black eyes of grays. They showed similar tiny mouths, ears, and nostrils to the standard small grays. No genitalia were visible. They had the same skinny legs and arms. Sightings and sources seem unreliable compared to a Philip Corso type of assessment. The Ebens seem like a short, minion gray but with smaller, more "normal" eyes.

REPTILIANS

Frightening reptilians, or Draconians or Dracos, about the same height of a man, appear frequently enough in abductee reports to be taken seriously. Scaly and strong are common descriptions. Their noses tend to be “snubbed” like a Gila monster rather than elongated, like an alligator. We can only make educated guesses about their position in the alien world and purpose.

A common claim is that they work with human military personnel.

Darwinism is not limited to Earth, and, like all species, they tend to act in their best genetic interest and proliferation of their identity and influence in the universe.

ALIEN GOALS

We must address the possible nature of the cultures and societies of alien visitors.

- 2004: Tic Tac craft tracked near the USS Princeton and USS Nimitz near San Diego. In addition to spectacular and overwhelming maneuvers, it showed it could decipher encrypted codes for rendezvous points.

- Assessed level of alien intelligence by Dr. Hartley Rutledge in the 70s and 80s: Genius and precognitive.
- If any living system could grasp, explore, and exploit parallel universes and interdimensionality, it would be the aliens.
- Aliens appear to operate at levels vastly superior to ours and, they are not from Kansas.
- Telepathic powers: nonlocal alien-human connections and telepathy in encounters are a universal given.

THE ARGUMENT FOR ALIEN SECRECY

Aliens do not want to disclose themselves fully. Based on the number of sightings and contact experiences in the records, many aliens didn't arrive until recently, say since the 1940s. They should have taken some time, replete with careful analysis, to assess our geopolitics and contact us formally. They would know we have weapons—and fears about creatures such as themselves. So, it would be downright dangerous for them to land and walk around like tourists. Continuous worldwide unrest and global bigotry and violence would slow them down.

We have a 10,000-year history, at a minimum, of the enslavement of our peers. We bear the burden of a history of beating, mutilating, enslaving, psychologically breaking, in imprisoning our kind. What would we do to another "intelligent" and highly skilled and competitive species? We stand accused of forming and spreading ideologies, including religions, often by force, and specifically fashioned to control the minds and behaviors of the masses. Digital media and its brainwashing ability are a modern adjunct to religion. For example, there may be exceptions, Tibetan Buddhism, but overall, mystic ideologies and most organized religions stand guilty as charged for mass mind and societal control.

In December of 2020, the former chief of Space Security for Israel for thirty years, Haim Eshed, made a public statement that world powers signed a peace treaty with a "Galactic Federation." He claimed the aliens did not want to alert the Earth about their presence because we could not fully understand space. But here is Eshed doing so? They think enough of us are ready to handle the truth, even though our governments do not.

No matter what course of action they take and how we initially respond, humans are likely to be a long-term threat to any other galactic civilization.

Alien secrecy may mean they are bent on controlling humans and/or their resources, including their biology.

The year 2020 is not the only year of an alleged treaty. Eisenhower met with aliens in 1954 at Edwards Air Force Base and got the worse part of a treaty—one that the aliens quickly broke—grays, of course.

Some galactic visitors may live quietly, or not so quietly, here on Earth under US government protection.

We can assume a form of competitive rivalry and hidden war is logical, predicted by Darwinian tendencies, and is highly disturbing. Are we but meat and treasure to be fought over and controlled? Probably. Instead, would we become a forthright and outstanding member of a galactic federation standing for peace, truth, and justice across the universe? Likely not! Captain Kirk, Dr. McCoy, and Spock we are not

Just one country might break from US influence and announce aliens independently: Iran. The problem is that Iran is not trusted and has no global media presence to stand up to Westerners' disinformation and ridicule campaigns.

THE ARGUMENT FOR SECRECY, PART II

We should address the secrecy issue for aliens enjoying a prolonged residence.

Since many of them evolved in radically different environments, aliens would have to re-engineer themselves to survive and propagate here in large numbers. If they just wanted samples of our life-forms and some rare minerals, there would be little need for communicating with us.

After thinking about the tremendous work involved in creating human-alien hybrids, or "hubrids," and sufficient infrastructure to support them—it simply became a challenging task for small numbers of aliens to achieve. A more practical approach would be to tweak humans in the direction of building sufficient infrastructure on their own—and then hybridize with them en masse.

Aliens may comply with interstellar policies, laws, or rules—or even interstellar AI police.

We may be more interesting, valuable, and possess much more potential to become a factor in the galaxy than we think.

TRANSHUMANISM ERA

A new type of human form, armed with modern technology and embedded in new forms of social orders and civilizations, becomes the transhumanist movement.

Powerful 5-G technology will erase all concepts of privacy from our psyches and catalyze a new world order, for better or worse. It may be mostly worse, a weapon applied by Big Tech to profit and exploit. Elon Musk-type neural-link devices that will link our minds to the internet, but also the internet to our minds. He makes a similar argument regarding super AI which eventually might run everything.

Genetically engineered babies happen underground and lead to a Brave New World caste system on top of the racial and wealth castes we know.

Aliens societies stand knee-deep in these modern technologies and have mastered their challenges and avalanches of forced change—we are the newbies on the block.

We can anticipate an acceleration in, as evidenced by the influence of the recent pandemic:

- Virtualization of society and global economies.
- Virtualization of shopping, education, recreation, healthcare, and sex.
- Complete loss of privacy.
- Digital money and banking.
- 5-G smart infrastructure.
- Increased restrictions on mobility.
- Increased censorship.
- Destruction of the middle class and independent farmers.
- Dominance by online business and personal transactions.

Alien societies are probably like this already: rationalized, unified, high tech, bureaucratized, and totalitarian.

Groups of aliens can be categorized as to their length of influence or physiology:

01 Ancients who got us started in the direction they wanted and stood back, in secret, and watched us develop.

02 20 th century arrivals, starting circa the 1940s, made appearances and meddled in wars and space programs. These would include the grays, mantids, and reptilians. Their agenda and methods: abduction, examination, extraction of sperm, ova, and other genetic material, techno-implants, and hybridization projects. Hybridization is continuous, organized, and widespread. Two main hybrid groups exist hybrids that will live with aliens and hybrids that will live with humans. Our CRISPR genetic engineering capabilities enable us to manipulate our genes. Imagine what the aliens can do. Subtle changes might be hard to discern, masking the hybrids for a long time.

03 Soul-implanted human-looking Avatars. Not a pleasant thought and impossible to detect.

WE FEAR THEM, BUT THEY MIGHT FEAR US

Considering our newfound technological prowess, especially if aided by an alliance with one or more alien groups embedded in a history of savagery, the rest of the aliens might fear our participation in the galactic family of civilizations stabilized by super AI.

Aliens' attempts to control and manipulate humans may just be in their self-interest in surviving our advancement, especially if we are yoked to an alien team of our own.

WHY EARTH?

Earth may just be a prime diamond in the rough. We have abundant and diverse life. We have an environmentally stabilizing moon. We will have a tech-supportive 5G soon.

WHAT IF THEY WANT TO HELP?

Nukes are a major issue for aliens here on Earth. They lecture abductees about the destruction of the environment, war, animal-like consciousness, and that more Obama-like "change" is coming. Do they care, or do they make arguments that will facilitate their taking over?

THE ULTIMATE COVERUP

We can cite the confessions of people high up in government regarding the undeniable existence of several types of aliens on Earth. Rather than the governments, the big corporations run the alien show (or no show) and tech circuses.

ADVANCED ANCIENTS

Some believe that aliens can form a human appearance in our minds if they don't have human bodies. He suspects any genetic enhancement, like a microcephalin gene enhancement, 40,000 years ago, would be their doing. Assimilating some humans into their society seems logical, and encouraging further advancement in intelligence also seems logical. Incorporating some aliens in human societies also seems logical.

The Ancients proved themselves peaceful and overall took a sit-back-and-watch approach. Continuous observation of humans for tens of thousands of years suggests that their society is controlled, rational, bureaucratized, and tightly organized by AI. The way the grays proved wired for communication from the Roswell crash means they may contact quantum AI 24/7 for their commands.

Francis Crick suggested Ancients engineered DNA, and it arrived here at least 3.5 billion years ago through panspermia, directed or not. The genetic code bears more resemblance to something that would arise from a computer and AI than natural selection. Panspermia by the Ancients might help account for many aliens taking humanoid shapes and may have to do with junk DNA pre-programming that could be activated from afar through electromagnetic waves to form humanoids when the timing was right.)

A GRAY/MANTID GROUP

Their unique appearance, set of behaviors, species psychology, and lack of sightings early on suggest that this group arrived on Earth in the 1940s amid WWII. This group demonstrates highly advanced science and technology, better than any human-looking group. They have no reservations about abducting people against their will and running experiments and hybridization projects. "Dangerous" is a fitting description. "Collaborating with governments" to influence our societal and physical evolution is another fitting description. "Mental manipulators" is another non-affectionate term.

Aliens would most likely demonstrate infiltration into leadership positions to mold policies in their favor.

REPTILIANS

Also alleged recent arrivals, and quite mysterious, technologically advanced, they seem capable of interdimensionality.

They seek to know how our minds and emotions work.

Just about everything Reptilian is conjecture based on rare encounters.

EBENS

These gray shorties represent more than one type. Their eyes are smaller than other grays, and their heads are different, demanding a distinction between groups.

They are suspected of abductions and collaboration with governments.

ONE OR MORE OTHER GROUPS

Large-eared shorties, carrot-ears, red eyes, or robotic are other types of galactic aliens reported. Reported data is sparse and inconclusive.

TRICKY DICKIES

Prescient intelligence of amazing capacities belongs to a special and advanced group. The subjects of spatial dimensions, time, consciousness, death, and afterlives permeate discussions regarding their presence and behavior.

Our knowledge of ourselves and the universe is, at this time, quite limited, and the best way to change that is through studying aliens.

FINAL THOUGHTS AND CONCLUSIONS REGARDING AGENDAS

As ex-President "income inequality is bad" Barrack Obama prepared for his 60 th birthday bash in his twelve million dollar mansion in Martha's Vineyards in August of 2021, one of three expensive mansions that he and his lovely wife own, planet Earth finds itself in an unprecedented crisis. Financial destruction looms, societies meltdown as crimes of violence escalate, the geopolitical summation can only be depicted with different degrees of insanity, and ways of life that lasted millennia find themselves on their death beds. "Change" occurred, and more is coming.

Our history of selfishness, greed, callousness, near-sightedness, violence, betrayal and the burden of emotional damage is not pretty.

We prove dangerous both to ourselves and any other civilization "out there."

The networks of 5G will facilitate a planetary totalitarian state under which all things Earthly will be micro-managed and steeped in advancements in genetic engineering and digital media.

We are being set for a leap, as a species, to a new level of existence that parallels alien civilizations. There is growing evidence for an ongoing alien intervention, yet there is no public discussion.

Do aliens seek to transform us so, as they hybridize with us so that they might control us and defend themselves against us?

Or is this standard Darwinian procedure, to encounter a new civilization and planet and conquer it, much like our colonialists?

Another irony: if it joins a galactic community, the human race will say goodbye to humanity.

We should recognize our biological beginning (even though it came from alien seeds) and our impressive display of initiative, energy, creativity, and tenacity to get us this far.

We need shots of more compassion as well as vaccines. Rather than the corrupt and greedy, we need the honest and the brave to step forward to lead. We need commitment to humanity. We need all of these to face our futures, survive, and put our John Hancock amongst the stars.

SECTION: POWERS

Given that some aliens may look very much like us according to abductees (FREE study) and that others might possess the ability to make themselves appear like humans, and we predict that DNA might organize many, to begin with, what might be the best way to distinguish them is through their demonstrated powers.

MOTION

With the discovery of a multitude of exoplanets, the chances of life elsewhere seem to be increased. Our ability to detect such planets continues to develop exponentially. When Nobel Laureate and discoverer of the structure of DNA Francis Crick first pondered life elsewhere, there was no scientific proof of planets in other solar systems—at all!

In terms of space and time and terms of worlds, the galaxy is huge, and so is the number of habitable planets with water and oxygen. This abundance has been underestimated, even by science fiction writers.

We now boast of evidence that our galaxy has five to six billion Earth-like planets, one billion of which prove themselves better suited to being blessed with life.[22] Five or six of these one billion can be expected to occur within fifty light-years, revealing them as our next-door galactic neighbors.

Assumptions of what is out there must change completely due to this new vision of the galaxy. Because worlds are expected to be more favorable for life than Earth, it is tempting to assume that complex life exists elsewhere, and if so, much of it probably emerged before ours. That does not mean that it had to originate on those planets!

Once intelligent life developed with sufficient technology, it could have spread itself systematically through a galaxy or even universe through directed--or undirected--panspermia. Directed means an advanced civilization spread it intentionally. Non-directed means a random chance process. One example is hitchhiking on comets or other heavenly bodies, perhaps still through engineered DNA. This most important theory and the supporting evidence are discussed at length in this book.

Even the Pentagon and the US Navy have officially stated that Earthly technology cannot explain certain observed flight patterns of UAP. They published videos of suspected UAP from "other" sources, implying non-Earthly ones.

This section lays out a basic theory for UAP propulsion that can help all observers to anticipate motions, behaviors. By studying the results of improved technological tracking, physicists now grasp how UAP work. For a compelling theory, the data is already sufficient and may lead to discoveries in physics.

OVERVIEW AND ASSUMPTIONS

The US Defense Department and Navy declassified three UAP short videos and revealed their sources.

A list of predictions based on the theory followed. The detailed data from the incidents can be analyzed and tested, pushing the entire study of UAP motion away from dismissals as swamp gas, weather balloons, and reflections and into the realm of the scientific method.

UNDERSTANDING PROTOCOLS

UA are tracked by several high-fidelity radar systems, infrared systems, sonar systems when they enter the sea, and personnel from carrier strike groups. According to all of the evidence, four malfunctions within a single system cannot occur across all platforms concurrently in the same way.

Based on this information, we can determine how these craft travel, their speed, and their apparent accelerations as they appear to us as observers. Modern radar systems offer 3D motion data to carrier strike groups to permeate entire battlespaces. The videos that have been officially released thus far comprise but a small portion of the Navy's total archives but are sufficient to form a sound theory. Visually perceived motions are significantly less informative than radar data on speed and maneuvers.

UAP display apparent linear accelerations, by Earthly observers and technology, to speeds faster than typical aircraft in less than a second. They can maneuver at acute angles as well.

During the year 2004, during the Tic Tac incident, the UAP went from 80,000 feet to various altitudes between 28,000 feet and fifty feet above sea level in 1.78 seconds

A motion in that period over a distance of even 60,000 feet would produce classical, Earthly-observed accelerations of over 12,000 g and speeds of over 100,000 mph.

UAP are routinely measured and seen to move very quickly, ranging from hundreds of g-forces to the extreme case stated above.

ONE IMPORTANT POINT ABOUT ET MOTION

The enormous apparent accelerations of UAP and sharp turns must be compatible with the laws of motion. These apparent accelerations are far too enormous and, at first, appear to break our laws of physics—as many misinformed claimed.

However, Einstein's published General Theory of Relativity paper includes a thorough analysis of space-time geodesics in which objects, and their occupants, can travel along with them and even accelerate without feeling any forces at all. This form of motion makes much more sense for UAP and explains their unusual movement.

Geodesics are straight lines on a curved surface that vary in shape depending on the curvature of space-time. Consider the space-time in which we live as a flat surface with no curvature caused by masses or energy placed in it. Geodesics, the routes taken if we free falling, exist in curved space-time. They guide all objects in space, including satellites, the moon, and the space station. Geodesics are the natural path through space-time when nothing pushes in any direction. This is why astronauts, free from acceleration, float about in near-zero gravity.

You can see why UAP have no accelerations as long as they move on geodesics. The UAP can navigate wherever they like, even at extreme speeds, because geodesics are fixed and based on the curvature of space-time. This will only work if the UAP employs a propulsion system that can construct geodesics by neatly twisting space-time into the proper shape. We might assume that powerful computers accomplish this, and even quantum computers seem indicated. This would allow the UAP to fall freely along any desired trajectory. If the propulsion system is efficient enough, UAP will not suffer from accelerations regardless of how fast they move.

The spacecraft “disappeared” from the pilot’s view during one encounter and then popped up sixty miles away.

THE THREE FAMOUS NAVY VIDEOS

GIMBAL AND ITS REMARKABLE HISTORICAL IMPORTANCE

The first of three US military videos, named "GIMBAL," tracks a stellar example of UAP. We benefited from the public release by the US government's formal declassification review process. This footage boasts critical chain-of-custody evidence. This certifies that it was received in its original, unmodified state. In other words, it is stamped as legitimate. The US Department of Defense uses this procedure to ensure that information and materials remain anonymous without disclosing their sources or techniques. This documentation distinguishes this footage from any other before entering the public domain, demonstrating its truthfulness, and endowing it with, in this case, great historical value.

While that fact alone is significant in history, what this 34-second video demonstrates is extraordinary. This one film contains several critical observations that may help us all gain a better understanding of quantum physics and technology at work. We need a quantum theory of gravity to describe—and "exploit" the universe completely. We need a quantum theory of consciousness to understand ourselves. (See later sections)

The aliens have mastered composing and applying the Theory of Everything or the God Equation as a symphonic map of the Universe.

Alien UAP might suggest how they use this technology to influence Humans and their environments. Examples would include not only powering up their spacecraft but also beaming objects into their craft, telepathy, cloning of technologically connected individuals like the small grays, or remotely disarming—or arming—of missiles.

AMONG THE SIGNIFICANT DISCOVERIES IN THE GIMBAL DATA SET ARE THE FOLLOWING:

• There are no discernible flight surfaces that are aerodynamic, mechanical devices allowing a pilot to adjust and control the aircraft's flight and altitude.

• Invisible in both the electro-optical and electromagnetic spectrums.

BRIEFING

The GIMBAL data results from one of the most advanced tracking systems ever installed on a US military aircraft. The GIMBAL craft found itself intercepted by a Raytheon AN/ASQ-228 Advanced Targeting Forward-Looking Infrared (ATFLIR) pod mounted on a US Navy F/A-18 Super Hornet. The ATFLIR is equipped with the most advanced sensors and tracking lasers available and can only be operated properly by an enlisted weapons sensor operator at the cost of millions for training. The ATFLIR has guaranteed high resolution and can locate and designate targets within a range of forty nautical miles.

The GIMBAL clip duplicates the amazing display in the cockpit for Super Hornet pilots.

The following are the primary characteristics and their implications:

VIDEO ANALYSIS

The Boeing FA-18-F Super Hornet pilots observe the following from their cockpit's console in a live video:

- The sensor operates in "white-hot" mode, which means the display's white parts are warmer than the dark or cooler areas. The UAP is represented by a white shape in the screen's center.
- The chasing aircraft is in a left turn at a speed of Mach 0.58 and is flying at the height of 25,010 feet.
- The UAP is going right to the left, a little below 2 degrees, and 54 degrees to the left of the Super Hornet.
- If we look attentively, we notice a black or opaque field surrounding or encasing the object.

Not only are the Super Hornet's pilots superb fighters, but they are also trained observers adept at evaluating their observations and potential targets to determine "friend or foe." They have been specifically taught to detect subtle changes in shape, size, position, flying altitude (angles), and speed to ascertain the threat's nature. They can notice information that other people would miss. Their ability to deal with stress while maintaining radio transmission becomes critical to their training. In the GIMBAL case, they are straining to comprehend what they are witnessing.

At 0:06, a different observer calmly adds, "There is an entire fleet of them."

Take a look at your ASA (radar display)."

"My God!" exclaims the first onlooker.

It is critical to notice that the ATFLIR displays a single item. Simultaneously, the radar provides the pilots with a view of the greater air area in which they are tracking several targets.

At 0:11, it is stated:

“They are all traveling against the wind.” The wind is blowing from the west at 120 knots (138 mph).”

As can be seen, the object’s speed and height are remarkable for what appears to be a drone-type aircraft. Based on this knowledge alone, the chance of an entire fleet of drones operating under this scenario is exceedingly doubtful.

Switched from “white-hot” to “black-hot” during this exchange, the sensors now show that the object’s image is considerably clearer. It is distinguished by its shape: a deformed oval with little protrusions at the top and bottom. The opaque aura of the thing has also become distinct: a “cool” glow that stretches a body thickness around the entire object. No discernible flight surfaces or exhaust plumes are evident, nor do there appear to be any of the standard components associated with conventional airplanes.

At 0:24, the object undergoes an abrupt change in height, indicating operating in a vacuum environment. Its direction and speed stay constant despite the 120-knot wind.

At 0:27, the object begins a sequence of distinct spins and shifts one hundred degrees in orientation. Despite the headwinds, its direction is now perpendicular to the horizontal plane. This movement violates conventional aerodynamic principles and further suggests a vacuum environment. As the film closes, the object’s altitude and performance appear to violate established physical laws, including those governing atmospheric resistance and conventional aerodynamic forces. It likewise comes to a near stop during the orientation shift but does not change altitude.

CONSEQUENCES

The famous GIMBAL video can now be formally classified as trustworthy, authentic “proof” of a UAP due to the chain-of-custody data. There is evidence of a flying vehicle with a shape that is typically associated with science fiction. No known technologies can be compared to what is observed in terms of performance and design, suggesting that a ship exhibits flying characteristics, unlike anything we know, understand, or can copy. We can conclude that the object applies technologies more advanced than our own because we cannot recreate them.

TECHNOLOGY SCREAMING "EXTRATERRESTRIAL!"

If any country, or aerospace corporation, had developed the technology to make possible the extraordinary motion that UAP show observers, it would have needed to create a complete theory of quantum gravity decades before the year 2004. Sightings of similar flights Tik Tac have been recorded for 250 years, beginning with the Industrial Revolution. [23] Ancient myths show themselves permeated with "sky fireballs" associated with unusual motion and with alien beings referred to as "skydivers" or "sky councils." [24] These myths date some 7,000 years ago back to Sumeria, which left us a couple hundred thousand cuneiform tablets as a record of life, beliefs, politics, myths, and culture.

BEYOND EINSTEIN

The needed equations, well beyond Einstein's theory of General Relativity, alone do not build such technology and craft as demonstrated by the Tic Tac UAP. They require insertion into physical systems for sound engineering and design. Therefore, adding on even more decades for development would seem logical.

Applications of such a complex theory would demand modern computing power, which simply did not exist, even remotely, decades ago. Aliens would use quantum computers for quantum calculations on the fly. Other innovative tools and technologies to manufacture suitable materials, like a ship material that can bend its shape in flight, like the one found at the Roswell crash site, would be one prerequisite to make a fully capable craft.

The lack of a complete theory of quantum gravity, not to mention the daunting engineering challenges, lack of appropriate materials and tools, would make it impossible for early technicians and engineers armed with slow and clumsy computers powered by floppy disks, which were first mass-produced in 1972, to make craft capable of UAP quantum gravity and entanglement calculations on the fly. Combining these facts and logical assumptions establishes with near certainty that the spacecraft technology is, indeed, extraterrestrial in origin.

The technology of UAP provides indisputable evidence that aliens visit Earth.

ANOTHER IMPORTANT POINT ABOUT ET MOTION

Instead of assuming that UAP were breaking the laws of physics, some physicists assumed they were obeying them. Some form of gravitational propulsion system is making these motions consistent with physics.

UAP must modify space-time to generate geodesics along which they can traverse.

Because gravity propulsion takes either too much energy or exotic kinds of energy, general relativity does not appear to allow it. However, because general relativity is not a quantum theory, it is incomplete.

When a new theory is proposed, it does not replace the existing ones; rather, it adds to them.

More detailed theories include quantum effects added to our understanding of the world – new developments that we may use to construct technology such as computers. Modern technology is based on quantum effects that are not present in Newtonian physics.

There is still a level beyond our most profound theories: quantum gravity, which is a quantum theory that describes space-time and its curvature or gravitation.

General relativity will be reproduced in this deeper theory, but with modifications and additions: quantum effects in space-time itself. If we already had this theory finished, we would take advantage of the new effects and use quantum mechanics to modify space-time and its curvature.

Quantum gravity does not completely elude physicists anymore. Even though there is no complete theory yet, physicists have more direction. Perhaps studying UAP will lead to a breakthrough. We are beginning to see how space operates at the quantum level with the work of string theorist Juan Martín Maldacena of Harvard and Princeton, arguably the leading theoretical physicist of modern times.

Physicists have known since 2009 that space can be described as a network of entangled quantum bits.

The link of all the points in space is the entanglement between the quantum bits. Space-time becomes but entangled quantum information. The degree of entanglement between two bits defines the spatial proximity of their corresponding positions. More entanglement means closer together. Less entanglement equals a greater separation.

SUMMARY OF NEW TERMS AND CONCEPTS FOR SPACE-TIME

We can redefine what space is, what distances in space are, and eventually, what constitutes motion based on quantum entanglement.

- Space-time is the entanglement of all quantum bits or qubits.

- Motions are changes in a qubit's entanglement.
- Gravitation is the curvature of space-time due to matter and harbors entropic components.

According to this theory and definitions, relativity may have a quantum correction.

Assume we had a comprehensive quantum gravity theory that would include a full description of the entanglement network. Like the aliens undoubtedly already have accomplished, we could develop technology based on quantum computers to manipulate space-time as a gravity propulsion mechanism. This would follow in the same way we can build quantum computers by understanding quantum mechanics.

Instead of thinking about it from a classical standpoint, you can now consider its quantum counterpart. We could manipulate the entanglement network, which is comparable to modifying space-time geometry, much more precisely and with much less energy than rearranging or altering masses.

This means that quantum gravity is applied to the hypothetical propulsion system of UAP in the same way that quantum mechanics is used to a quantum computer. In general relativity, we know of geometries that move at extremely high speeds. Therefore, this technology could even enable interstellar travel.

PREDICTIONS

For want of a better phrase, requiring that UAP be consistent with the laws of motion leads to a gravitational propulsion system, which has the following proposed characteristics: The following is based upon the information available at UAPTHEORY.COM and other sources.

- Because there is no preferred forward-facing direction, UAP should be round or geometrical. They don't need wings or afterburners since they employ space-time as their control surface. We were able to designate the known designs of UAP because we demanded that they obey the laws of physics in this universe.

- Particles move around UAP, avoiding colliding with the craft by following the curved geometry of the area around them. As a result, UAP have a vacuum envelope around them and can move freely in water, air, and likely other media. They can enter and exit the water without resistance. They can move more quickly than any craft that is in contact with the air or water.

- Because a vacuum envelope surrounds it, UAP may appear to be encircled by glass. Light travels from the air into the vacuum, then back into the air. When light travels from the point of view to a vacuum, it is refracted and partially reflected, just like when light travels through glass or other material. Direct sunlight may be lensed, reflected, and refracted, especially at high altitudes, resulting in sparkling patches or rings around the UAP. A pilot may get the erroneous impression that the UAP is encased in glass.

- Unlike aircraft, UAP move very quietly or silently. When they go supersonic, they might not make a sonic boom.

- Because UAP move on geodesics and are in zero-G, they can maneuver at sharp angles at high speeds.
- We should detect distortions surrounding them when they change directions since they have to adjust the geometry of their field. We should see the UAP change shape as it accelerates or turns if this optical effect is large enough. The background of the UAP may likewise be altered in its vicinity. Light paths follow the curvature of space-time as they enter and escape the distortion field, exiting at different angles. If you try to perceive a UAP through its distortion field — as an object in a distorted glass container — you will never see it directly. Suppose the UAP wishes to maneuver or modify its speed. In that case, it must alter the local distortion of space-time to construct an alternative geodesic, which will cause it to fall in a different direction. Because the light routes change with the geometry of the field, we should see the object change shape. UAP should change their apparent shape when they change direction or fall along a new geodesic. It's worth noting that we can only expect this while they maneuver or accelerate, as they may continue to move at a constant speed after that. A distortion field isn't required for constant-speed motion. This means that we may still see a UAP mostly undistorted while it maneuvers, but we must watch for changes in its shape.
- UAP may be invisible from below due to aliens selecting a distortion field that causes light falling from the ship to be redirected back up or to the sides. If such a cloak is operational, we should observe one or more UAP mirages from those higher vantage points. If a UAP intends to emit a beam, scan the environment, map the ground, or attack, it cannot be disguised and must seem undistorted.
- Otherwise, any beam would be redirected away from the intended target.

- UAP should have a high energy output, causing the air to heat up and have a distinct thermal signature. To prevent discovery, a cloaked UAP may run on low energy and speed. If someone gets too close to it, it might inflict burns.

- Before fading, some UAP should rotate prominently. We expect the UAP to employ all of its space-time manipulating potential fall at its top speed because any craft's maximum speed requires its engine to work at full capacity. However, the machine's primary function is to hold the object against Earth's gravity and perform maneuvers, which we assume would necessitate a specific ship orientation based on its construction, pointing the engine down or up. To get from typical movements to high speed, the item would have to rotate towards the journey point for the machine to focus in that direction. Spacetime is compressed in front of the vessel and expanded behind it in this geometry. We predict a spin of the craft before a jump to high speeds, at least with some UAP designs, particularly flat designs. Field generators can be deployed all around the shape of the vehicle. Therefore, spherical, or cube-shaped UAP may not be constrained by this requirement.

BLOCKING MEMORIES

Abductees repeatedly report some form of memory loss from encounters. We must study the theories of memory and determine how aliens may be blocking memory.

Proceedings of the National Academy of Science contains a research project that indicates that deliberately targeting a specific memory, or declarative memory, can degrade it.

Moreover, no medications, electroshock, or other invasive methods were utilized to erase memories. Research on laboratory animals shows that memory may be tampered with and subject to mistakes and confusion through timely manipulation of protein synthesis in the brain.

When neurologists Jessica LaPaglia and Jason Chan first presented their findings at a scientific symposium, they opened a bag full of mixed reactions. "Some enjoyed it a lot, while others didn't," he remarked. "I couldn't believe it myself when we got the data three years ago." Both working at Iowa State University, Chan and LaPaglia replicated the studies and tested the results in additional settings. They held up.

The researchers had erased a specific memory by interfering precisely at the right time in the multi-phase biological process.

"What's nice about his study is that they employed a memory that's extremely comparable to real-life memories (a movie in this case)," said memory expert Daniela Schiller. They also showed that the approach only worked when initiated at precise memory storage and recall phases rather than inducing global amnesia. To suppress memory, she added, it must be reactivated during the interference.

A memory element dubbed "reconsolidation" has been identified in laboratory animal studies. Building long-lasting connections between neurons requires the continual synthesis of new proteins. If you stop protein synthesis shortly after learning anything new, the experience won't be remembered long-term. But timing is crucial. Wait too long to block protein synthesis, usually a few hours after training, and the memory is permanently stored. [25]

ELECTROMAGNETIC FIELDS AND MEMORIES

Even if each electromagnetic field oscillator in the brain is weak, the sync between them can create a strong and very selective force on memory.

György Buszaki made his position on this argument known as early as 2004, pointing out that the brain's electromagnetic fields link diverse brain sections together, facilitate synaptic alterations, and create and consolidate memory. So Buzsaki concedes that these fields are like a beating heart for the brain. [26]

QUANTUM CONSCIOUSNESS, SPACETIME, AND THE ORCH-OR MODEL.

Nobel laureate physicist Roger Penrose and anesthesiologist Stuart Hameroff, MD, are the "go-to" researchers and experts on quantum consciousness models.

OLD STANDARD MODEL

The assumption for the standard (or old) model l] consciousness arises from complicated computation among brain neurons, with neuronal firings ('spikes') and synaptic transmissions equating to binary 'bits.' Consciousness is thought to have 'emerged' from intricate neural computation as an adaptation of living systems, extrinsic to the universe's makeup. Conversely, certain spiritual and contemplative traditions and scientists and philosophers regard awareness to be intrinsic. Platonic forms and conscious precursors existed before biology in the fine-scale structure of space-time reality.

ORCH-OR MODEL: ORCH IS "ORCHESTRATED," "OR" IS "OBJECTIVE REDUCTION"

Dr. Hameroff's research posits a theory of consciousness developed over 20 years with prominent British physicist Sir Roger Penrose. Their Orch OR theory proposes that consciousness starts from quantum vibrations in microtubules, which interfere, 'collapse,' and resonate through the brain. They influence neuronal firing, which, in turn, forms consciousness, Microtubules with their quantum communication establish connections to 'deeper order' ripples in spacetime geometry. To Penrose and Hameroff, consciousness becomes more like music than rigid computations. [27] [28] [29]

Recall that Colonel Corso in The Day After Roswell [30], given that the little gray and cloned aliens had chip implants in their brains, experienced further electromagnetic synchronization with their ship through their suits, and wore designated pilot headgear, all to help them achieve the ability to guide their craft with their minds. We see similarities between Corso's statements and the nature of bending spacetime as the basis of their motion to Penrose and Hameroff's statements about the microtubule connection to "deeper order ripples in spacetime geometry."

Consciousness, according to the 'Orch OR' theory, results from quantum computation in microtubules. Microtubules organize ('Orch') alternative collective dipole oscillations--quantum bits--to process information. These coordinated oscillations entangle, compute, and collapse in sequences of Orch OR moments with choreographed conscious experience. Orch OR selects microtubule states that control neuronal activities and offers a testable theory for how consciousness works.

Therefore, consciousness is attributed to quantum computations in microtubules inside brain neurons (Hameroff & Penrose, 1996a, 2014). Entanglement unifies quantum bits or qubits that evolve and compute until reduction, or 'collapse,' to definite output states. Owing to an objective threshold in the fine-scale structure of the cosmos, Nobel Laureate Sir Roger Penrose (1989) hypothesized that reduction occurs spontaneously at a time defined by an equation with the Planck-Dirac constant and the gravitational self-energy of superposition. [31]

The field of Neural Quantology is controversial but attracted a number of publications connecting holographic and fractal models to early universal consciousness sensitive to light and sound.[32]

ALIEN TECHNIQUE FOR SELECTIVE MEMORY WIPING AND TELEPATHY

Anesthetics may selectively wipe consciousness clean by affecting interactions at the quantum level inside microtubules (Emerson et al.[33].) Specifically dampened by anesthetics, sophisticated computer modeling reveals collective terahertz oscillations amidst microtubulin's 86 pi electron resonance rings (Craddock et al.[34].)

We know that aliens mastered the concepts of creating changes in spacetime to traverse the universe along geodesics. Might they also master the same concepts to understand Human consciousness at the quantum level and above and apply their knowledge to erase memories in Humans selectively, or to render them completely unconscious—perhaps by using some form of quantum interference?

Might this also underlie what seems to be their efficient quantum –and perhaps entanglement-based telepathy, including the projection of emotions, images, and entire memories rather than wiping them?

CONCLUSION AS STATED BY STUART HAMEROFF, MD.

Orch OR is the most comprehensive explanation of consciousness, spanning fields and scales, and has significant explanatory power. However, if quantum interference in microtubules ('Orch') cannot be demonstrated, or if it can be established but is found to be insensitive to anesthesia, the theory of Orch OR will be disproved and discarded. Orch OR is the most comprehensive and easily falsifiable theory of consciousness currently available. It has not been falsified yet.

Orch Or helps explain how aliens might apply quantum interference or coherence, or quantum entanglement or disentanglement, to influence Human consciousness.

Many memories of abductees arose from applications of hypnotic regression, which seems to release the contents of the more unconscious workings of the brain. [35] Likely more of an amorphous amalgam rather than selected pathways, it's probably more difficult for aliens to block.

Learning how the aliens manipulate our memory so well reflects a broad understanding of our consciousness, and understanding their exact techniques may open up entirely new fields of medicine.

A breakthrough in the biological sciences involved studies at the quantum level. When scientists mapped mathematical descriptions of single-cell processes and the branches that arose from them and how they interact with one another — anatomically and functionally — they got a very similar description of quantum events. Complex biological processes, at their most minute levels, tend to follow the principles of quantum mechanics. We may never understand ourselves unless we completely understand quantum mechanics. [36]

Aliens mastered quantum mechanics, and hence they know us, including our consciousness, better than we do. In studying aliens and their skill sets, we may ultimately know ourselves.

WHICH TECHNIQUES WOULD ALIENS USE TO WIPE HUMAN MEMORY?

Applying quantum interference seems like the most efficient way for aliens to wipe memories based on quantum activity in the microtubules in the human brain since they are so well-schooled in quantum mechanics and perhaps even quantum gravitation.

Aliens might use a combination of quantum interference with chemical anesthesia to render someone unconscious before an abduction and perhaps during and after, depending on intent and durations.

Some abductees claim they were given a lecture through images about the fate of the Earth, for example. The abductee would have to be conscious to get the point.

If the intent is to perform surgical procedures that don't demand a conscious state, a number of sophistical chemicals might be used. The aliens would have to be careful about blood samples after returning the abductee showing trace residues of drugs and giving Humans information about procedures.

Killing neurons would undoubtedly have negative side effects.

Manipulating and selectively disrupting the brain's electromagnetic fields during attempted recall seems to be a good possibility since the aliens have mastered quantum mechanics and spacetime.

HYPNOSIS

We must consider hypnosis for recalling perceived experiences with aliens and as a means for aliens to repress the abduction memory.

POWERS: TELEPATHY

Telepathy is the ability to read minds and communicate through the power of thought alone.

A list of applications is nearly endless. The CIA and the Soviet Union funded and trained psychic and remote viewing teams during the Cold War, and the experiments continue today.

Perhaps the most exciting and well-known research comes from a team in Barcelona that works at Star Lab who recently have been able to send a word, without saying it, though directly from a human brain in India to a human brain in France

Scientists use two main methods, nudging closer to genuine telepathy over the years. The first method is to tap straight into the nervous system by inserting wireless communication chips into the appendages of the sender and receiver. Brain messages are transmitted, received, and through muscle movement. Right now, Humans don't have a chip small enough to be implanted straight into the brain—but I would bet some Aliens do.

The current implants are so big that they can't attach to a specific neuron when placed inside the arm. They must settle for reading many neurons together, making it more challenging to send a noiseless signal.

The second method bypasses implants altogether, replacing them with an EEG cap that channels brain signals directly to computers by reading the electrical activity. You can then transmit the brain pattern to anything with a good receiver.

The researchers associate the readings with actions, movements, or words.

Researchers can only read very simple patterns in the brain that translate into about five messages. A current research team combined both methods, motor movement through muscles and brain EEG signals. The subjects thought of the word "bird" as a message but first translated it into a binary signal, by on and off signals created by clenching and unclenching their fists. This signal is kind of like Morse code could then be transmitted from India via computer to a human receiver in Barcelona who rigged up with a transcranial magnetic stimulation device. The device created on-off flashes in their brain, which could then be translated back to binary and finally decoded back to the original word, a rather indirect route, but still telepathy. The clinching of fists became a form of simultaneous sign language mixed into the process. Nevertheless, it worked! Making all components faster, more specific, and wireless would get the scientists closer to pure telepathy.

There are other examples of successful telepathy experiments:

- Moving a rat's tail remotely by a Human mind.
- Using EKG helmets to send and receive words by a different team, with no appendage movement message.
- Predicting who was to call on the telephone
- Showing how a monkey can play a computer game with its mind at Neuralink.

Some people think that quantum entanglement could underlie telepathy. [41]

In addition to the ORCH-OR model, researcher Gao Shan proposed a quantum model of consciousness that includes telepathy. [42]

Since quantum fields form the fabric of the universe, it would come as no surprise that they would underlie the evolution of life consciousness and remote communication such as telepathy.

GMASTERS OF TELEPATHY

How might aliens practice telepathy? Entanglement unifies quantum bits or qubits that evolve and compute until reduction, or 'collapse,' to definite output states through the microtubules in our brains. We might expect aliens to have something similar or a mechanical duplicate?.

Suppose we could describe a memory, emotion, or perception as an evolving set of qubits at the quantum level. In that case, it seems reading it from an entangled collection of microtubules and neurons in another brain should be feasible. This transmission and reception could be enhanced through quantum AI and quantum computers. The message might be directed in some fashion or transmitted into space but received only by entangled minds or machines. This type of "guess" could be developed into a testable hypothesis.

The Devil Towers mountain image was projected to Earthlings as a future meeting place, but only certain people responded. Our hero, Richard Dreyfus—who had connected with the aliens, in other words, had become entangled with them and became so obsessed with the image that built replicas of it in his living room and his mashed potatoes.

For aliens who mastered shaping spacetime and entanglement for traversing geodesics to account for their motion and likely all of quantum mechanics and quantum gravity, mastering telepathy would be accomplished with the same skillset.

POWERS: PRECISION SURGERY

Corso worked on a laser from the Roswell ship that could be held in one hand, and his consultants concluded it served as a precise cutting instrument that could be used for surgery or other tasks. The famous cattle mutilations that began in the 1940s seemed to be one application, showing extremely precise and narrow diameter cuts that Human instruments could not duplicate.

POWERS: IMPLANTS

Some abductees claim they were implanted with mechanical devices or saw others receiving implants on alien ships. Sometimes they found themselves injected into the brain through the nasal passages. [43] They were assumed to be tracking devices or technology that might facilitate the telepathic reading of a person's mind and activities.

If any had been recovered, surely the government would confiscate them and censor any witnesses.

We would assume that the aliens would be highly proficient in designing and implementing nanotechnology for many purposes.

Corso claims that the little gray alien clones had chip implants and became electronically with their ship.

POWERS: CLONES, CYBORGS, AND BIO SLAVES

In The Day After Roswell, Colonel Corso explicitly stated the government's conclusion about the small gray aliens found in the Roswell craft lived as EBEs or Extraterrestrial Biological Entities. They did not refer to them as "aliens" or "beings."

Because of implanted chips, we would also label them as cyborgs. They arose from genetic engineering, cloning, metallurgy, and computer implants. This conclusion emerged from the five autopsy reports that Corso had the clearance, and the assignment, to review.

They served as dedicated slaves for space travel. They did not eat, drink liquids, have any digestive system internally, and had little in terms of a nose or mouth, if at all. They seemed to live in the craft indefinitely, perhaps absorbing any required nutrients through their skin and spun-on suits.

Their design allowed for electromagnetic resonance with their ship and its piloting systems. Corso referred to it as "aligned atoms."

Speculation: A combination of their superiors and quantum AI likely guided many of their activities and behaviors, probably driven by a model of quantum consciousness. They lived not as individuals but part of a "hive-mind." Perhaps they lived for hundreds of years in a small craft, making them a cyborg bio slave. They were probably cloned to form a massive population

POWERS: DISARMING MISSILES

Ex-military leaders in the United States allege that aliens tampered with nuclear weapons. At least four former American Air Force captains have come out in the United States to allege that extraterrestrials are responsible for a number of events that have occurred over the last few decades. They are ready and willing to go "on the record" with their allegations that weapons systems have been tampered with, making them unusable, if necessary.

MALMSTROM AIR FORCE BASE AND OTHERS

Air Force captain Robert Salas says that he realized his whole weapons system, consisting of ten intercontinental ballistic missiles, became inoperable. He was stationed at Malmstrom Air Force Base in Montana, where he was the on-duty commander on that particular shift on March 24, 1967. Malmstrom Air Force Base is home to nuclear weapons, and he was the on-duty commander on that specific shift.

He says that his system even initiated a launch sequence but could not accomplish anything since his controls were not working. Fortunately, the launch countdown was stopped before it reached zero minutes. He further says that a similar incident had occurred at another ICBM site eight days earlier, on March 16, 1967, at which he was present.

They wanted to hold a news conference on October 19, 2020, at the National Press Club in Washington, DC, with the help of four former United States Air Force captains. They were willing to talk publicly about their encounters with suspected extraterrestrial and UFO influence after collecting £10,000 to cover the costs of the event's expenses. They also claimed that they would make declassified government papers available to the public.

The four military men held this news conference in the hopes of persuading the United States Congress to undertake a thorough inquiry into UFOs. They also requested that a public hearing be held on the events at the nuclear power plants. Additional sightings were reported later above the Buckley Space Force Base in Colorado when a homeowner recorded odd objects in the sky and uploaded the clip to social media. [44]

HOW DO ALIENS DISARM MISSILES?

Aliens demonstrated that they needed to get close to missiles or control panels to disarm them by a low-altitude flyover. They likely use electromagnetic fields and quantum AI to breach security and hack into missile computers to disarm them. In all cases involving missiles, the UAP had to execute a flyover at close range, perhaps to increase entanglement with their targets through increased proximately and increase the precision of their acts of disarmament or armament.

BEAM ME UP, SCOTTY

Some abductees report traveling up into an alien craft in mid-air. Some report seeing beams projected from alien craft.

How might the aliens accomplishing "beaming?"

We can speculate that the aliens beam objects by applying similar technology they use to travel along geodesics. For an object, they just adjust spacetime, so the object "falls" upward into the ship. They might also apply fields from the side to ensure the object does not traverse the lateral boundaries of the beam.

Colonel Corso knew of the aliens lifting a missile off of a truck bed with such beams.

SECTION: DNA

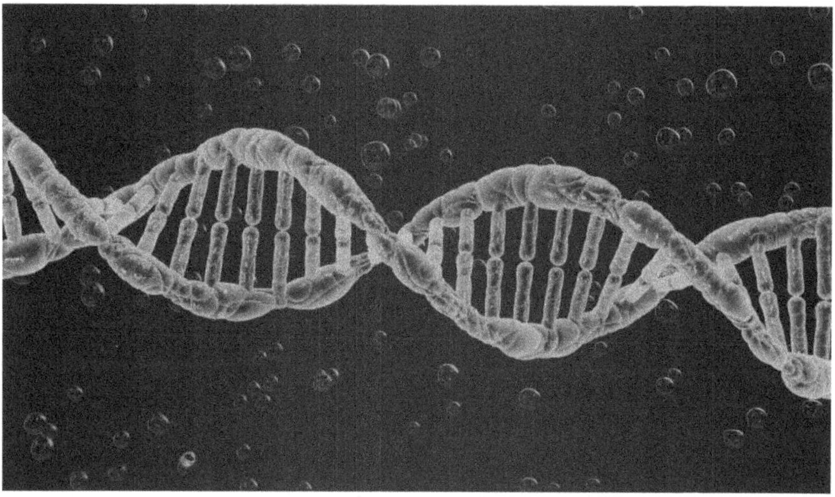

Figure 3 DNA

DNA IS NOT FROM EARTH

The most profound questions revolve around our origins. Unfortunately, answers, often illogical and without evidence, still embarrass us as self-serving myths.

Although it is not yet common knowledge because of blatant censorship, how we came about is now very clear, as supported by two distinct lines of scientific evidence. The answer remained hidden in our DNA all along until discovered in 2013. DNA, it turns out, is a masterpiece of intelligent biomathematical coding by an extremely advanced alien civilization, making all life on Earth of alien origin. The probability that we had alien stewards is ten-trillion to one. [45] Further intervention may have boosted our intelligence.

The Exogenesis Theory, also known as the Directed Panspermia Theory, or simply the Seed Theory, became first popularized by Nobel Prize winner and co-discoverer of DNA Francis Crick. The theory further begs biological and evolutionary proof of alien "super-microbes" on Earth. The proof presents itself in our family of Giant Viruses and our microbe evolution, leading to all lifeforms.

"Mitochondrial Eve," about 200,000 years ago, the single mother of all of us, brought forth Homo sapiens, which boasted a robust increase in the size of skull and brain and a new, elongated shape. It proved itself to be about 150 cubic centimeters bigger than those of modern humans. Although the brain only is about two percent of our body mass, it consumes about twenty percent of our energy. Improved mitochondria for increased brain energy became part of the Homo sapiens package. Perhaps aliens had intervened. There are other times in history when radical and unique changes occurred in very short evolutionary time spans for primates, for example, about seven million years ago, when pre-humans split off from the rest of the Great Apes. [46]

It wasn't until the middle of the 19 th century that Charles Darwin published On the Origin of Species in 1859, suggesting that life evolved rather than an act of creation by an unknowable God. Finally, science offered a viable and testable alternative explanation. Evidence for evolution occurred everywhere, but the origination of life still posed a mystery—the greatest mystery of all. We seem to know much more about how our universe started with the Big Bang than we do about exactly how the first life came about on Earth.

OK, life evolves, but how did it all start? Darwin once suggested that some little warm pond near the ocean was a spot. Genes would not be discovered until seven years later by Mendel and DNA a few years later by Swiss chemist Friedrich Miescher.

Francis Crick's book titled Life Itself promoted the overwhelming and scientifically logical idea that DNA did not first evolve on Earth. DNA's alien origin would be one of the most important realizations in all history. Life, according to Crick, began somewhere else where conditions were more favorable and, with the help of advanced intelligence, seeded Earth.

If DNA originated elsewhere, it reflects our beginnings as a form of "alien" or "alien hybrid" and our future, including merging with one or more advanced civilizations. Indeed, we may meet our "makers," or at least some of their descendants one day.

Crick's insight and intuition found themselves confirmed by convincing scientific evidence from a prominent mathematician and astrophysicist. They published a paper in

2013 in a peer-reviewed journal named Icarus, formerly edited by Carl Sagan. Icarus just happens to be the same journal that published Crick's paper over three decades earlier elaborating on the possibilities of Directed Panspermia, also known as "exogenesis," by advanced civilizations.

When one of the co-discoverers of the structure of DNA claims it is of interstellar origin, we have to listen to why he believes it is so. The evidence to support his claims is mounting. First, DNA is an elegant mathematical and signature construct--which means it is artificial. Second, the Last Universal Common Ancestor, or LUCA, appears to be newly discovered bizarre giant viruses chock full of genes unknown to Earth. Third, the United States government reveals increasingly about UAP sightings and materials associated with them and says they cannot rule out an alien presence.

We now have the evidence of signatures and hidden mathematical and pictorial codes from DNA for Crick's theory. We update cosmological data that Crick did not have at his disposal, address the subject of the surprising Last Universal Common Ancestor for life on Earth, which has alien origins. We benefit from our government's about-face on the issue of UAP and admitting they study bizarre materials found at their landings. Taking a close look at videos and data released by the United States Navy of UAP reveals flight patterns arising from technology based upon mastering theories of quantum gravity and entanglement that have eluded Earthlings, even the efforts for thirty years by Einstein.

If aliens visit Earth, what exactly do they want and do, or did in the past? Some claim that many of our advances in technology arose from reverse-engineering the technology of aliens. Some believe that advanced civilizations would want to form hybrids with us. Others think that they want to conquer and enslave us.

NOBEL PRIZE

In 1962, Francis Crick shared the Nobel Prize in Physiology and Medicine for the remarkable 1953 discovery of DNA's molecular structure, a task that evaded other prominent scientists in the race, such as Linus Pauling.

Crick stood as one of the most knowledgeable scientists in the world concerning the structure and function of DNA but surprised everyone with a controversial and overwhelming conclusion that DNA proved to be too complex, too well-organized, too logical, too powerful, and just too perfect to be fashioned from natural selection alone. Furthermore, he stated that it did not likely originate on Earth but arose from genetic engineering from alien civilization billions of years ago. DNA, according to Crick, reached Earth through a process called Directed Panspermia.

Stirring up a potpourri of questions about our very existence and the nature of life, Crick's elaboration on Directed Panspermia bears implications for life's origins on Earth and humanity's future. First, he addresses the probability of intelligent life elsewhere. Second, he sparks a debate about whether we, in turn, should be directed panspermists trying to spread our version of the genetic code about the universe.

Crick imagined unmanned rockets with cargoes of primitive spores organized into life by DNA that included planet Earth as a potential landing and seeding site. DNA could also have hitched a ride on such heavenly bodies as meteors, asteroids, or comets. For example, the famous Murchison meteor that hit Australia in 1969 had dozens of organic compounds, including amino acids, and revealed its age at over seven billion years.

Survival of the journey, impacting Earth, and staying alive and evolving on the surface are all addressed by Crick. He describes the Big Bang and the nature and expanse of the universe in determining an estimate of the chances that life exists elsewhere.

If Crick proves correct, then the same ancient alien civilization or civilizations could have spawned life on many planets. These life forms on other planets may be our brothers, sisters, and cousins. Perhaps timely visits by the engineers to those planets that had an appropriate atmosphere bore life and tweak life further towards advanced intelligence in preparation for eventually increased hybridization would be a good Darwinian strategy to spread one's civilization and genes armed with viable adaptation strategies already in place.

One disturbing possibility for pure humanists and religious folk alike, not considered by Crick, is that DNA arose from alien artificial intelligence's design and propagation efforts.

WHO'S YOUR DADDY?

The Big Bang exploded about 13.8 billion years ago. One hundred million to a billion years later, stars, planets, and other heavenly bodies began to populate our universe. z Let's estimate on the conservative side since we have all of the time in the world in this case and say that it took a billion years. Intelligent life evolved on Earth in about four billion years, even given that our DNA arose elsewhere. On a planet with a history of more hospitable conditions, it might have happened sooner. Again, let's estimate conservatively and say it took five billion years. Let's assume that our universe is fourteen billion years old, and it requires a billion years to get habitable planets, and assume that it took five billion years after that to evolve intelligent life capable of interstellar travel.

SPECULATION

If it didn't perish, the first advanced civilization could be as old as eight billion years, or even up to ten billion. When it was four to six billion years old, or much sooner, such a civilization could have genetically engineered DNA suitable for space travel and colonization through the microbes associated with Francis Crick's Directed Panspermia. About four billion years ago, it could have colonized Earth. Science is now telling us that Earth became a recipient of genetically engineered DNA from somewhere . It could have been from an advanced civilization that became impressive genetic engineers from four billion years ago to eight billion years ago.

We could argue that Earth was not the only planet that would benefit from DNA seeding. Such seeds could have a planet to themselves or compete or merge with any existing life on many worlds. Suppose we project Darwinian theory onto aliens and assume that they would spread their genes and identity as far as possible as a survival mechanism. In that case, Another possibility is that the "ancients aliens" observed the life they had cultivated and tweaked along the way to serve their needs, plan, and goals—and still do.

Another possibility is they had and have territorial competition from other species throughout our universe—including Earth.

Imagine technological advancement for four to eight billion years in one or more civilizations. What would they be capable of doing and achieving?

If we project more Darwinian urges, including diversification, onto such civilizations, they will have many engineered physical and interdimensional forms. Many would-be infused with sophisticated quantum artificial intelligence. They would have some intelligent robots, androids, and cyborgs.

Would they be peaceful? Let's hope so, or we might find ourselves enslaved. We might also conjecture they are fully capable of defending themselves to have lasted so long.

This hypothesis is based on the biological and non-biological, namely the mathematical characteristics of DNA.

Therefore, it is testable.

Comparative DNA analysis between Humans and aliens would validate it. Further analysis of our DNA might hint at ancient alien components or additional mathematical insight and communications.

Assume ancient alien civilizations could boast eight billion years or more in our universe. Think of their capabilities and adaptive forms, undoubtedly limitless. It would be natural for many of us "primates," wired for spiritualism and religiosity at the expense of truth and logic, to think of them as "gods." We want to follow them, be healed by them, and learn from them. Religions operate amidst fantasy and delusion, but the advanced alien phenomenon offers up actual, physical, or interdimensional "gods" as advertised by their technology.

Much of Crick's book finds itself devoted to exploring Fermi's statements of steps to advancement as a template against which all theories must be compared.

At the time of Crick's book, there proved to be no hard evidence that other stars hosted planets at all. Now we have a proof for estimates of trillions of planets across multiple galaxies.

Crick argued that we couldn't accurately estimate the time frames to go from soup to interstellar spaceships. Some of the steps were painstakingly slow, which would help explain why no alien visitors stood in line at Disneyland.

The truth may be that we have been visited numerous times already. Many believe a UFO crashed at Roswell in 1947 during a lightning storm, which had been announced over the radio by the US government and stood to be later confirmed by many witnesses. The story became retracted the next day as a mere weather balloon. "Sightings" of UFOs and descriptions of aliens are in the historical records for over five thousand years, beginning with Sumeria. Some cave paintings, tens of thousands of years before that, depict alien influence. Many military personnel reported sightings, and particularly near nuclear missile facilities. Consider the US government's release of videos that show spectacular Unidentified Aerial Phenomena. Considering that the physics of their flight patterns belies a thorough proficiency in quantum gravity and entanglement, which still perplexes even our best physicists, they are unlikely to be human creations, although reverse-engineering would be an exception. The US government admitting that they possess materials suspected of being "otherworldly" is more evidence. Just take a good look at all humans on Earth and, in fact, all Earthly lifeforms. Tell Fermi's admirers that, if Crick and other panspermists are correct, they only need to look in the mirror to see evidence of an alien lifeform on Earth.

Crick and cohort Leslie Orgel instead proposed in a paper that the living spores arrived in the head of a spaceship, which was unmanned, pointed towards Earth, and similar ones towards other planets, by advanced civilization billions of years old. Life, they reasoned, began in the ocean, which would be the place to make a landing and have the most ingredients of a primordial soup. (Modern theory suggests underground is a distinct possibility, too.)

They labeled this concept as Directed Panspermia. Ironically, their article earned a spot in the Icarus space journal, a periodical edited by Carl Sagan, the same journal which would publish astrophysicist Vladimir I. shCherbaka's and mathematician Maxim A. Makukov's paper claiming an alien signature and coding, including decimalization and an expression for zero and activation codes, to our DNA over three decades later in 2013, more evidence that Crick's hunch proved correct. A spaceship wasn't necessary to make the trip or transfer if we assume our DNA cosmonauts could live for billions of years in solid rock, as some endoliths, indeed, promise to do.

THE GENERAL NATURE OF LIFE

Crick is amazed at the high degree of organized complexity that we see at every level in the living organism, especially at the levels of molecules. Proteins impress Crick the most, and even the simplest ones may be built from over two thousand atoms which must maintain their relative organization with each other for the protein to function in its role.

Nature reproduces proteins by amazing sleight of hand. They are synthesized as extended one-dimensional structures and then fold themselves up. The folding process proceeds under the direction of the precise patterns of the side-chains, in constant interaction with each other and with the backbone with the help of multiple weak interactions. The best possible fold is facilitated by thermal movement, and the parts neatly join up together, now protecting their creation from unwanted additional thermal activity.

Molecular construction, a minor miracle, proves to be achieved by stringing together amino acids in proper order to make polypeptide chains through the help of messenger RNA. Amazingly, 10^{260} proteins can arise from coding involving just twenty amino acids.

DNA, ELECTROMAGNETIC FIELDS, AND QUANTUM TELEPORTATION

Teleportation is the rapid movement of matter from one location to another.

Luc Antoine Montagnier and his science colleagues have all been awarded the Nobel Prize in Medicine for demonstrating spacetime teleportation of DNA information. His team included two prominent theoretical physicists.

Quantum teleportation (QT)) transfers quantum information from one quantum system to another using entanglement. It cannot transfer the system itself. Rearranging macroscopic object particles to duplicate another thing's shape is also not covered. It can convey information in a quantum superposition, valuable for quantum communication and computation. [47] [48]

No matter or energy is transferred. Only the state of quantum particles is passed. This indistinguishability of quantum mechanical particles is a form of "teleportation." [49]

Montagnier's team "teleported" the DNA sample. A DNA polymerase was used to realign the polymerase to a copy of the DNA sample. It's like rearranging water and alphabetic characters into a comprehensible text.

A DNA polymerase is an enzyme that polymerizes deoxyribonucleotides into strands of DNA. DNA polymerases are most recognized for their feedback function in DNA replication, where they use an intact DNA strand as a template to synthesize a new strand. This process copies DNA. That molecule is complementary to the template strand but not identical.

We need the "organizing power" of this DNA sample to shift the single "alphabetic characters" to their exact placements or enable its reassembling by adjusting the probability that it occurs. [50]

THE PRIMITIVE EARTH

What would be the primordial soup ingredients sufficient to kickstart life? We find carbon, hydrogen, oxygen, nitrogen, and less sulfur and phosphorus in all life forms. Given recombination potential, less than fifty different atoms allow for an extremely high number of other macromolecules.

Where did these essential atoms for organic compounds and the other atoms we need for chemical reactions like magnesium, sodium ions, potassium, iron, and chloride, come from?

Organic chemistry plays favorites with highly reactive atoms. Crick believes RNA may have come first and precipitated simple forms of protein synthesis. But, of course, if living spores arrived from space, such as blue-green algae, or a giant virus, the machinery would all already be present. It would simply suck up what it needed as raw materials from the environment, the ocean, or underground thermal vents.

Without the benefit of panspermia, Crick argues that RNA would first arise from random-based sequences, contradicting what we see today and early on in the first evidence of life. Experiments do not support the idea of success with just RNA and t-RNA alone, that is, without protein or ribosomes.

Crick explores numerous possibilities of kickstarting life on Earth without panspermia and comes up short. He concludes that it is impossible to imagine how a restricted number of components in a little pond or somewhere underground could have formed the first life on primitive Earth. The source of precursors is also a complete mystery. No scientist can boast success in showing how salts, water, UV light, and a handful of gases came to form an RNA replicating system. The time measured from when Earth first became hospitable to the necessary compounds to the start of life itself is a mere three hundred million years, a small increment of time in a 13.8 billion-year-old universe. Crick informs us that it is a problem figuring out how an accurate system in replication arose from the complex mixture, which is believed to have birthed life. It is even more challenging to determine the exact components and steps required. We have no idea how RNA replication started, much less how it became coupled to protein synthesis.

Crick concludes that it seems impossible to assign a numerical value to the probability of starting life on Earth, obviously an unlikely sequence of events. However, Crick rejects arguments that life inevitably had to occur and instead concludes that life is extremely difficult to kickstart on its own. So many, says Crick, are the primitive conditions necessary that the start of life seems quite miraculous. If this is true, Crick asserts that we must consider that life first arose not on Earth but where conditions were more favorable on an alien planet.

A STATISTICAL FALLACY

Crick points out that just because life is here on Earth, it is a high probability that it arose elsewhere as well or would start again here if we had to do it all over again, which Crick refers to as a statistical fallacy. We have no factual basis to reason that, although the probability of starting life from scratch may be extremely small, the billions of planets with hospitable conditions make up for it, and the beginning of life in many places is inevitable. The Human mental weakness of superstition, generalizing from a single instance, rather than scientific fact and logic, is behind this view. Hunters and gatherers applied generalizations, quick and emotional, to survive. We succumb to another Human weakness when pondering life's origins. It is difficult for us primates to wrap our minds around large numbers and distances. However, Crick considers the possibilities of planets where conditions for birthing life are much more favorable.

WHAT WOULD THEY HAVE SENT?

Crick unleashes his healthy imagination and postulates that intelligence arose on another planet that mastered technology and science far beyond our current understanding four billion years or so ago. Crick gives us another thousand years to show impressive advancement. He gives us another 10,000 years to be pretty remarkable, and particularly in engineering projects.

He postulates that they would have acted if they concluded that many other planets had good primordial soups for life to begin but occurred without a viable chemical mechanism sufficient for natural selection. Suppose knowing that their planet's sun would eventually die, or just from Darwinian urges to spread one's genes and identity, would be enough motivation to transport said chemical mechanisms for natural selection to enough planets to trigger new life.

Microbes, such as endoliths which thrive in rocks, or protected giant viruses, would be the most logical candidates to send out in space in all directions in unmanned rockets or attached to meteors or comets in an attempt at colonization. Bacteria may live with little oxygen around and, in fact, with fermentation, no oxygen. They can survive on little to "eat," have mobility, and live for extended periods under extreme conditions, including intense radiation like that found in nuclear reactors. Space is full of harmful radiation. They reproduce asexually, which is not overly exciting but extremely efficient and convenient. Some are photosynthetic. They employ cell walls as means of protection against a great many hazards and threats. Our earliest microbial fossils came from blue-green algae, one of the best candidates to be early cosmonauts. Some live in oceans, some in freshwater. Hot springs have blue-green algae, so do deserts. Crevices in rocks? Blue-green algae will be there.

Bacteria are great candidates because they don't take up much space, and they can be frozen and unfrozen like nothing happened. When frozen, they can lay dormant for extended periods. Even just at low temperatures, they can live up to 10,000 years. So, impact shock would not bother them, nor a variety of other hazards that would do in mice, chimps, or Humans.

If they landed in an ocean filled with nutrients, it would suit them simply fine.

A bacterium that could act like a virus or become one easily to infect and destroy competing life and replace them with forms based on alien DNA would be an excellent choice for those aliens who wished to conquer. The giant viruses fit this description. They offer the best of two worlds, the living bacteria and the potential "non-living," but a highly infective virus.

THE DESIGN OF THE ROCKET

An extremely fast rocket arising from advanced technology, which traveled close to the speed of light, would operate under time dilation. Time would pass much slower for its astronauts than for the inhabitants of the planet from which they came. From the perspective of one home sun to another, a journey to colonize another planet might take thousands of years, but time for the astronauts might only burn up a few decades. Cryogenic sleep might make such a trip more bearable. The rocket also must resist the hazards of space that might damage it, no easy task because of the high speeds at which particles might impact the spaceship.

Studies of UAP conclude that they modulate spacetime near them to place themselves on new geodesics to change directions and speed but proceed under no accelerations. Distortions fields embed the craft when it changes direction or speed and creates a vacuum between the craft and the media it travels through. Therefore, such craft would not have to worry much about "particles" impacting their ship. The proposed technology allows for traveling faster than the speed of light, which challenges, of course, Crick's early arguments about constraining effects from the long timespans regarding space travel.

Crick concludes that sending out humanoid astronauts to colonize other planets with bacterial life may not have been practical even for very advanced civilizations. Instead, he reasons that sending unmanned ships would make much more sense, with microbe-laden payloads of about 200 pounds, even though they would have to survive a trip for up to ten thousand years and work well enough at the end to employ an optimal strategy to seed an entire planet for life with multiple forms of bacteria, ranging from the very primitive, like blue-green algae, to advanced eukaryotes like yeast which can survive with or without oxygen.

Crick argues that even a very advanced civilization might employ microbes to spread life to planets in other galaxies, which would require extremely long travel times. Microbes, he says, can go farther. Unfortunately, the concepts of using wormholes to travel vast distances in shorter times, as suggested by people like Stephen Hawking and Michio Kaku, were not prevalent in Crick's time because he failed to mention those possibilities at all. Nor was propulsion by proficiency in quantum gravity and entanglement mentioned.

THE TWO THEORIES CONTRASTED

Theory One: Life started on Earth independently with extraordinarily little or no help outside our solar system. Theory Two: Directed Panspermia: Life started on another planet, reached forms that included advanced intelligence capable of impressive engineering projects, including sending genetically engineered bacteria out as little cosmonauts to colonize other planets, like Earth.

Crick begs the question:

" Does it matter?"

Crick points out that since any form of life occurring early on in the universe after the Big Bang would be impossible, life had to start at least a billion years after the Big Bang to have all of the necessary ingredients available.

Compared to Earth, life could have formed much earlier and evolved more rapidly. Earth didn't even begin for the first nine billion years or so after the Big Bang. Other planets may have had a huge head start eight billion years. The fact is that the same genetic code, except for mitochondria, proves to be virtually identical in all life on Earth.

Mitochondria are different but not radically different.

The fact is that all of our genetic code dictated by DNA adheres to a purely chemical logic. The coding hints that life had to go through a bottleneck at some point, with a tiny interbreeding population producing all subsequent energy as their descendants.

Crick reasons that at least a few versions of DNA should have arisen and persisted. But they did not. We only have one, aside from the slight variation of mitochondria which formed a symbiotic relationship with cells. This lends credence that the miracle coding of DNA underlying all life on Earth came from the skies above through efforts of Directed Panspermia by alien engineers and geneticists.

Crick also asks if there is any other feature that all living things have in common on Earth. Early life shows several distinct families that seem like distant cousins, which is also weakly supportive of the theory of Directed Panspermia. Blue-green algae did occur early, but we don't have enough knowledge to say how significant this is for our history.

FERMI'S QUESTION RECONSIDERED

Crick informs us that some people believe there is absolutely no evidence of alien visitation to Earth, and therefore we must be the only higher intelligence in our galaxy. They believe any advanced civilization would have already spread itself about our universe. Consequently, one does not exist.

Crick reasons it might take as little as one million years to populate the Milky Way, or up to one hundred million depending on conditions and technology. Surely they would have encountered Earth by now. Of course, this argument depends on assuming that the smarty pants aliens would want us to know that we were being visited. Rather than hide them or try to destroy them, our governments would welcome them and introduce them to the populace.

Colonizing planets is no easy task since it requires great feats of engineering once the adventurers arrive. Oxygen might have to be injected into the atmosphere or harvested from the labors of bacteria. Cultivation might require major terraforming. Surviving as a colony for any considerable length of time might not be so easy. Any one of many catastrophes might doom them to oblivion.

Directed Panspermia would make much more sense, and the facilitating technology to be successful would come much earlier in the history of technological innovation. Remember, bacteria can go farther, much farther, can live much longer, survive many forms of extremes in conditions, and would prove infinitely more cost-effective. Sending out bacteria to build up a proper oxygen atmosphere suitable for future generations. Many scientists propose that we do it to Mars, and ancient aliens did it to Earth.

Crick discusses the possibility that alien civilizations are sending us signals, different attempts at interstellar handshakes. The 'Wow!' signal received by SETI in 1977 in Ohio seemed so unnatural and came across so loud that it might have, indeed, been from an advanced alien civilization. [51]

Crick also addresses the inherent challenges of successful signal transmission.

He concludes by suggesting that figuring out if we are alone or not in our galaxy may be a matter of our survival.

WHY SHOULD WE CARE?

Crick argues that demonstrating the creation of life from a laboratory chemical soup would unify lives and bind people closer to Nature.

He also insists that science works to uncover answers to our most profound mysteries and therefore deserves an aura of urgency. Rather than superstition and religion, science pursues the truth and offers the best explanations for it.

The questions about the origins of life immediately give rise to questions about possible distant neighbors in our universe, and how to send signals, and even the meaning of life in the first place.

EXPANSION: WOW!

In May of 2013, scientists Vladimir shCherbak and Maxim Makukov published a paper entitled The "Wow! Signal" of the Terrestrial Genetic Code in which they suggested that DNA was proven to be embedded with a variety of sophisticated coding and activation keys which includes mathematical constructs, not the least of which are the use of decimals and zero. DNA, they say, first arose from artificial engineering, agreeing with Francis Crick, and bears cleverly encoded mathematical and pictorial signatures of its makers.

Makukov and shCherbak explain the technical challenges of an intelligent civilization transmitting messages at optical or radio wavelengths that would persist, undisturbed, for lengthy durations of time, say millions of years. Terrestrial life's genetic code demonstrates a flexible mapping between amino acids and codons that permits artificial code modification. They showed how a genetic code could carry intelligent messages or signals and perform all requisite biological functions. Living cells can preserve the genetic code, in theory, for billions of years and carry coded intelligent messages.

Recall that Francis Crick, a co-discoverer of DNA structure and one of the greatest scientific minds of the 20th century, believed DNA did not evolve by natural selection. In other words, Crick convinced himself that DNA not only did not originate on Earth but also arose from talented genetic engineers who dispersed it into space for directed panspermia.

After thirteen years of work concerning the genome project, Makukov and shCherbak provided solid evidence for Crick's hypothesis by discovering multiple types of coding embedded in DNA that have an extremely low probability of arising naturally—one in ten trillion or so. Their results imply an intelligent alien designer who left a cosmic John Hancock.

DNA-driven life on Earth arose from alien genetic engineers in a panspermic seeding from non-Earthly sources. Sources could have been meteorites, comets, or planetary fragments, or colonizer spaceships.

Repetition of the number 37, which occurs nine times in the code, suggests an alien signature.

According to shCherbak and Makukov, DNA appears to be a mathematically perfect creation meaning it could not have arisen naturally.

According to the scientists, the configurations of the genetic code reveal a set of ideographic and architectural patterns of symbolic language. It features decimal notation, a zero representation, and systematic and precise, and logical transformations. It features constants, fractions, and mathematical proportions.

DNA also uses symbols that represent physical objects or ideography. Some of the symbols are apparent universal symbols of life.

shCherbak and Makukov believe that DNA contains a form of a receiver for electromagnetic waves.

We can speculate that whoever genetically engineered DNA may know how to influence the performance or function through electromagnetic waves.

Humans, like all life on Earth, arose from alien sources. Interstellar microbes contributed heavily to our evolution, and we leave the issue open as to whether aliens modified early humans into becoming Homo sapiens.

If we assume that even if the first DNA arrived on Earth by a "natural" carrier like a meteorite, that said DNA still must have sprung from advanced bioengineering because of its mathematical codes and signatures. It appears that at least one civilization evolved highly advanced intelligence while we were still in a primordial womb.

Information stored in DNA's genetic code is, in fact, the most durable known to Humankind, and maintaining its fidelity for billions of years is an automatic given.

Microbes, especially those adapted to living in rocks, like endoliths, would be among candidates capable of interstellar travel. However, as tested at the International Space Station, we have extremophile bacteria on Earth that survive in open space. [52]

Humans evolved through DNA, which we propose arose from another planet, with Homo sapiens emerging 200,000 years ago. A spike in brain size appeared some seven million years ago and again, 25,000 to 60,000 years ago.

Did the designers and spreaders of carbon-based DNA follow up and monitor those planets where life flourished? Did they want sufficient infrastructure to support their lifeforms on multiple worlds after invading them in some fashion?

Did they want to spread their genes, artificial intelligence, and identity to as many planets as possible, a primal Darwinian urge, and survival insurance policy? Would forming hybrids with Humans and other life on Earth be one way of achieving that? Would exploiting at least early Humans make sense to build the desired infrastructure and perpetuate their lineage?

Wouldn't we do the same?

Would other alien species realize that the work of the original DNA engineers would offer an opportunity to bootleg planets full of life for their purposes? Is there alien competition or cooperation throughout the galaxies?

Do wars and alliances exist throughout the universe?

Do the unpredictability and violent and warring nature of Humans and their historical use of nuclear weapons serve as cause for alarm for all alien civilizations?

Given that nuclear weapons may negatively affect space-time to disturb the well-being of interdimensional entities or space and time travel, are we a species to be controlled absolutely or exterminated?

The number of UFO and UAP sightings escalated dramatically in the middle of the 20th century after using nuclear weapons in the 1940s. Nuclear weapon facilities are common for UFO sightings and defensive acts of sabotage and disarming by suspected aliens, as reported by military personnel from multiple countries.

MORE EVIDENCE

Vladimir shCherbak and Maxim Makukov maintain that Nature is indifferent to languages based on numbers and designed by intelligence to depict zero and other quantities. They point out that a privileged number system becomes a reliable sign that Nature is not responsible for it—the system is an artificial one. Suppose such a number system occurs as intentionally embedded in an object, such as DNA structure. In that case, such a privileged system might demonstrate itself through a distinctive notation to any viewer carefully and systematically examining that object. In other words, “decoding it.”

Vladimir shCherbak and Maxim Makukov continue by stating that, to be concluded unambiguously as a sign of intelligence, all code patterns must meet two criteria:

001 They must register as highly statistically significant.

002 Not only must they display intelligent-like features [53], but they should not reflect, in principle, any natural process, various pre-translational mechanisms, genomic fluctuations, driven by amino acid biosynthesis, affinities between amino acids and (anti)codons, selection for the increased diversity of proteins, or the energetics of codon-anticodon interactions.

Vladimir shCherbak and Maxim Makukov show the calculations of the statistical significance of the coding they found in their second appendix, Appendix B. There is only a probability of one in ten trillion that the coding arose naturally.

Because there might be unknown natural processes, the second criterion may be unverifiable because the patterns, they say, could result from such processes. They argue that this criterion is the same as asking if it is possible to embed practices representing specific information into the code to earn an unequivocal interpretation as representative of an intelligent signature. The answer, they claim, is yes. The two scientists suggest that one way to do so is to make patterns virtual, not actual, and is precisely what is found in the genetic code. Even more impressive is that strict balances and decimal syntax become detectable only after employing the "activation key."

Remarkably, t he signal patterns are first projected as a system of algebraic expressions, which are then solved straightforwardly to reveal the mapping of the code.

These signals in the genetic code serve as evidence of a signature of intelligence. The signals must also represent contents that bear logic. This is why the code of DNA underlies such an abundance of life shielded from possible ill effects from chance modifications. If reassignment of a codon arises from a mutation in tRNA or aaRS, all genes in the genome remain transcripted with the prior code. A meaning of irrelevance engulfs the mutated cell rapidly because the mutation could not replicate.

Because of the presence of dozens of codons in a genome comprised of genes that number in the thousands, the probability that mutations in aaRS or tRNA in one codon coincide with similar mutations in all genes is minuscule.

The genetic code machinery demonstrates tremendously robust purifying selection, keeping it unalterable for potentially billions of years.

There is no other medium known to Humankind that can match this type of unalterable duration, making it a fantastic carrier of communications across vast stretches of space.

The nucleons of amino acids and the bases of nucleic acids seem most suitable for digitization. The researchers compared the genomes of Euplotes octocarinatus and E scherichia coli. Repeatedly, the nucleon sums of amino acid standard blocks and side chains were discovered to be exactly matched. Moreover, the balanced sums' digital notations acquired the unique decimal form 111, 222...., 999. This form results from the 037 divisibility criteria. The criterion could simplify a cell's computing mechanism and speed up calculation. The genetic code's cooperative symmetry shows that this mechanism was designed and used a zero. The activity of hypothesized molecular organelles could explain this genetic code structure. [54]

ANCESTORS OF THE FIRST GIANT VIRUSES AMONG THE FIRST ALIEN LIFEFORMS TO COLONIZE EARTH

The ancient ancestors of viruses who stand as giants in comparative size in the virus universe initiated complex life on Earth as LUCA, the Last Universal Common Ancestor to Archaea, bacteria, and eukaryotes.

Giant Viruses, up to thirty times bigger than most other viruses, provided the raw material and needed symbiosis for the evolution of cellular life and assisted in driving its diversification into the astounding myriad of organisms that permeate our planet. Giant viruses can be as big as some bacteria and hold as many genes as possible. [55]

Although the Giant Viruses share most of the genes found in all life, many of their genes are unknown on planet Earth. Some giants even have genes that are involved in DNA replication. They have genes that occur in no other known life forms on Earth. Some produce energy.

At the National Institutes of Health, a computational biologist, Eugene Koonin, believes in the Virus World theory that viruses came first in the primordial soup, and complex life evolved from them. Koonin published a paper sharing that sentiment in June of 2014 in Microbiology and Molecular Biology Reviews. [56]

Another Koonin, Victor Koonin, published a paper in 2006 in Research in Microbiology with Christos A. Ouzounis, Nikos Darzentas, and Leon Goldovsky that described the application of an algorithm for ancestral state inference of gene content, which compensated for gene loss, which predicted that the Last Universal Common Ancestor (LUCA) of the three major domains of life maintained over 1,000 gene families. They characterized, functionally, 90% of them. Considering only prokaryotes, the number ranged between 1,006 and 1,189 gene families. When eukaryotes are added, they found between 1,344 and 1,529 families, depending upon which part of the tree of life they examined.

Computational genomics of the 21st century did not support the common belief of the 20th century that LUCA should be much more primitive with the small genomes like those in pure parasites. Rather, they discovered a much larger and complex genome reminiscent of the genomes found in free-living prokaryotes. They reasoned that the LUCA possessed various functional capabilities, including information processing, metabolic transformation, membrane and transport proteins, and complex and shared regulation between the three domains of life. [57]

Giant viruses also encode complex metabolic capabilities but have evolutionary histories which prove to be mostly independent of cellular life. This fact makes us consider them as significant drivers of biogeochemical cycles. [58] Some giant viruses can metabolize for energy and have remnant machinery to process information to replicate RNA and DNA and make proteins. They also demonstrate membranes. [59]

Mimivirus encodes mitochondrial machinery making it a candidate for the first source of mitochondria on planet Earth which then spread through parasitic injection beginning \some one and one-half billion years ago.

These possessions, abilities, and the size of their genomes qualify the giant viruses or their ancestors to be among the candidates for LUCA.

BUILDING BLOCKS

Jack Szostak , a Nobel laureate and biochemist at Harvard University, gathers evidence that viruses or their ancestors were integral to some of the most important stages in the evolution of complex life, including the emergence of DNA and the birth of the first cells. Thus, the tree of life's early root branches may all reflect interactions with early viruses. To align with Francis Crick's exogenesis theory, among the first cosmonauts stood ancestors of our giant viruses.

Life is transmuted into multiple major domains: bacteria, eukaryotes, and Archaea. Some scientists argue that giant viruses should be among them to make four major domains of life. Unicellular organisms bacteria and Archaea merged into multicellular eukaryotes.

Viruses are among Earth's designated "bad boys" because we associate them with disease and pandemics. According to modern theories, viruses either evolved after a jailbreak or from "degenerates." Viruses might have come from genes that managed to escape the "jails" of cells or from genes in degenerate cells that had given up on replication.

The big bad boys, the giant viruses, hid from scientists up to 2003. Then, totally unlike other viruses, they bopped scientists over the head in 2003 to the degree that many started to rethink our entire evolutionary history. The giants, it seemed, looked like a form of a missing link. Giants proved unique and had tricked some scientists into thinking they were strange bacteria, earning the name " Mimivirus " previously for Mimicking Microbe Virus. Mimivirus has mitochondria, which is otherwise unheard of in viruses. Most giant viruses approximate the size of a bacterium, stain like a bacterium, and have as many genes as a bacterium. However, to virus experts equipped with electron microscopes, "Mimi" showed the tell-tale twenty triangular sides of a typical virus, albeit extra-large ones. The stargate structure, characteristic to giant viruses and adapted (or intelligently designed?) for injecting proteins into cells, seen on the surfaces of these viruses, attracts scientists "owing to its beauty and symmetry," as well as the fact that no smaller viruses have one. Mimivirus even produces energy in-house. It's hard not to call Mimivirus "alive."

After their first discovery in 2003, virus hunters Jean-Michel Claverie and Chantal Abergel began to find more and roped and caged a huge and mysterious giant living in rather unlucky amoebae--some giants turn victims literally into a stone-like material after killing them--off of the coast of Chile. They would appropriately label the beast as Pandoravirus because of the "box of mysteries" it held inside that now the scientific world could see and ponder.

Boasting over 2,500 genes, Pandoravirus deserved its branch classification. Moreover, more than 90 percent of its genes did not occur in any other Earthly lifeform. Thus, Pandoravirus genes were deserving of the term "alien" and fit the description of something we might expect from another, older planet.

The discovering scientists reasoned that giant viruses were ancient, and parts of their genomes must occur ubiquitously even though the giants had remained hidden to scientists because of their surprising enormity of size. Nobody thought to look for an overgrown virus that mimicked bacteria.

Within a few more months, the virologist headhunters netted yet another and labeled it Pithovirus -- which proved to be even larger than Pandoravirus and had many unknown genes to Earth and with anonymous functions.

The disturbingly abnormal genes led to speculation about the beginnings of giant viruses and life itself. Moreover, their abnormal genes jolted scientists, forcing them to consider the possibility that the ancestors of giant viruses had occurred extremely early in the history of life on Earth, even in the primordial soup itself.

To make the absolute best case for the giants to have been among the lifeforms that won out to compete for LUCA, and for being the descendants of alien cosmonauts capable of panspermia, we need to embrace the "Virus World" hypothesis that says viruses not only came before complex life but assumed multiple roles to become integral to the very evolution of all complex lifeforms.

Although not comforting to some, the evidence suggests that Humans evolved from parasitic viruses filled with alien genes--that is, mini-aliens. Strange giant viruses claim that their alien ancestors beat out the competition to become the precursors to all life on Earth and probably many other planets, as well.

THE VIRUS WORLD

The Virus World theory states that the ancestors of modern viruses came into being when all potential life was still a floating smorgasbord of mostly lipids and amino acids. According to the theory, RNA formed the first bits of genetic material in tiny pieces, comprising just a few genes, but that was found guilty of parasitizing its neighbors. Thus, consuming its neighbors, some of which were also made of genetic material, allowing for replication. These primitive amalgams of genetic information exchanged and, through natural selection, effectively used some new acquisitions and discarded others.

According to this theory, life first evolved from primitive forms of parasites. Interpreting this with an exogenesis perspective, advanced alien civilizations bombarded Earth over four billion years ago with genetically engineered or manipulated microbes, including viruses and giant viruses. Ancestors of our giant viruses that could replicate DNA or parasitize others would be a logical candidate—a conquering alien beast, of sorts.

Given enough time, some primitive parasites became dependent upon others for replication, evolving into viruses that exploit hosts for replication and other needs. The host genes evolved, in turn, means for protection, including new forms of genetic information, and formed the blueprints for the first cells.

The RNA World Theory, a close relative to the Virus World Theory, claims life first arose as tiny bits of RNA that eventually evolved into complex organisms that employed DNA for replication. The Virus World Theory concurs with the RNA World Theory in that life's genetic material started as a form of RNA and claims that virus ancestors evolved before cells.

EVIDENCE

Diversity belies the onset of origin. Diversity reflects age. The earlier an organism arose, the more diverse it should become. The more descendants an organism has, the earlier it must have appeared. Part of the evidence for Human life in Africa is that its residents show much more genetic diversity than anywhere else on Earth.

Viruses show much more diversity than their cellular counterparts, suggesting they came first and much earlier. Science and evolution must follow the rules. If it works for Humans, it must also likely work for viruses.

If cells came first, they should be more diverse.

They are not. Viruses are far more diverse and came first.

If cells came first, they should hold the entire spectrum of genes that occur in viruses.

They do not. Viruses came first.

In contrast, the giant viruses demonstrate genes from a wide variety of organisms. All of them! This suggests they acted as an early "mother" gene donor for other life forms in addition to evolving more diversity.

When we take a look at modes of reproduction, viruses show much more diversity. Replication of DNA occurs in only two ways in cells. One way came from bacteria, or an ancestor of the giant viruses since they have some DNA factory machinery left over, and the other came from Archaea and eukaryotes.

The classic three domains of life, bacteria, Archea, and eukaryotes, share some of the same viral proteins, suggesting that those viral proteins evolved before life diversified. Adding a fourth domain of the giant viruses would place the giants earlier on the tree of life than the other three domains.

Some researchers, like Patrick Forterre, believe viruses lived before the last universal common ancestor (LUCA) of all life on Earth. [60] Some believe the last universal common ancestor may very well have been a giant virus. [61]

If LUCA lived behind the mask of a giant virus, seeding all life on Earth, why would many of its genes not occur at all in *any* known Earthly lifeforms? We would expect this if LUCA, synthetic or not, came from another planet.

Strange as it seems, given the immensity and thoroughness of the Genome Project and gene identification of all species, many of the giant virus genes are unique or "alien." Consider the bizarre giant virus known as *Pithovirus sibericum*. Fully 66 percent of its genome reveals itself as dissimilar from the genomes of the *Pandoraviruses* and other known genomes on Earth. The surprising presence and genomic composition of giant viruses such as *Pithovirus sibericum* suggest that LUCA must be defined as the bifurcation point of the Eucaryotes and Archaea from the giant viruses. [62]

Let's assume that Francis Crick, Vladimir I. shCherbaka, and Maxim A. Makukov are correct; DNA arrived through exogenesis. In that case, giant viruses harboring many unknown genes are readily explained. These novel genes represent life on another planet or synthetic concoctions and are not, and will not be, found in other Earthly organisms. Given that the giant viruses are the main candidate for LUCA, the giant viruses or their ancestors might have carried alien DNA that they continued to replicate.

An alien origin would also account for why the giants are so strange and unique in many ways. Bacteria tend to be like bacteria, and viruses tend to be like viruses, but giant viruses share characteristics of both bacteria and viruses.

Since they share many characteristics with bacteria, they could have made the interstellar journey as bacteria, and then a portion of them lost that identity through degeneration. The giants arose from genetic engineering to be infective and parasitic, so they could eventually replicate their alien DNA in hosts and spread themselves rapidly to seize the LUCA title eventually.

If today's scientists from Earth were to engineer a bacteria-sized organism to infect another planet with our DNA and compete with other lifeforms, then an organism that could either infect a host cell and use the host's machinery to replicate, or replicate on its own, as the giants did at least at one time, makes perfect sense as a viable genetic strategy for conquest from a single type of organism.

ALIVE OR ZOMBIE PARASITE?

Traditional viruses are often not considered a form of life because they miss the means to replicate their genes and instead exploit their hosts, which could be cells, algae, or amoebae, to reproduce.

Giant viruses find themselves in a phylogenic "alive or not alive" limbo. However, giant viruses still carry some of the genes involved in DNA replication, suggesting their ancestors may have once existed as free-living organisms with nuclei.

Some scientists are adamant about giving the giant viruses what they deserve: their branch on the early part of the tree of life. The giants could carve their name on a branch just below bacteria, Archaea, and eucaryotes and not disturb them—other than suggest they had a hand in their evolution. An official giant virus branch would pay tribute to the genetic freakishness of the giants with genes, and lots of them, that do not occur elsewhere on the Earthly tree of life. Recent research based on "protein folds" confirms that giants came early in life's history rather than later. [63]

Are they Earth's original ancient aliens?

Despite harboring numerous genes that occur nowhere else on the planet, giant viruses have been mistakenly lumped into a bigger family called the "nucleocytoplasmic large DNA viruses. This group includes the smallpox virus. Giant viruses demonstrate much more complexity than smallpox viruses, and scientists initially thought they evolved later than their more traditional viral cousins. They were bigger and more complex. More recent—and accurate-- work confirms that these viruses evolved exceedingly early in the history of life.

In a published study in 2012 at the University of Illinois, Gustavo Caetano-Anolles , a bioinformatics specialist, investigated the giants' protein history. Concerning LUCA, the giants either predated it or at least coexisted with it. If the giants prove to be as old as Caetano-Anolles claimed through his calculations, over two billion years, the implications are nothing less than overwhelming. To tie the theories together for an alien exogenesis, the ancestors of giant viruses became the top dogs in our primordial soup. The shocking takeaway: a giant virus, if we consider them "alive," or one of its ancestors, not only lived before other types of life. But played a highly significant role in evolving all life on Earth. This could mean that the giants and other viruses should be viewed as one of the most powerful evolutionary forces on this planet and that each Earthly organism carries a viral past. Earth was—and still is—a Virus World. Recall those remnants of DNA replication machinery in some giants, suggesting that their ancestors made their DNA simultaneously, unlike other viruses.

According to Dolja, cells required genetic gifts from viruses to have evolved at all. To go from RNA to DNA requires reverse transcriptase, an enzyme." In modern times, only viruses have it. However, giant viruses still have some of the machinery to make DNA, so perhaps they had an equivalent of reverse transcriptase at one time, and they would have to be a candidate to have ancestors who were the first DNA-carrying cosmonauts

Abergel and Claverie claim that giant viruses evolved from a line of cells that is now extinct. Giants, they find, live everywhere. Ancestors of giant viruses went "pure viral" and gave up their ability to replicate—and their independence—to exploit host cells instead. Bits and pieces of cellular remnants of the giants survive in modern Pandoravirus, Mimivirus, and Pithovirus , which would help explain at least some of the genes found in this group that occur in no other lifeform.

Life was not, and never will be, simple. Claverie claims that life participated in a fight Darwinian brawl with many cell-like organisms competing rather than one ancestor. The winner of the competition, LUCA, took all and birthed all complex life on Earth.

LUCA is looking increasingly like a giant virus or giant virus ancestor as new scientific data comes in to inform us.

Jean-Michel Claverie and Chantal Abergel did not address the possibility that giant viruses/bacteria-laden with alien forms of DNA, and an abundance of non-Earthly genes, landed in the primordial soup some two or more billion years ago after a panspermic trip through interstellar space to catalyze complex multicellular life on Earth —but we are.

It would be intriguing to hear their opinion on the matter.

Some viruses lend themselves to crystallization like minerals when they are in their nonliving state, which supports surviving for millions of years until they find themselves in moisture and a host. The dormant viruses may regain activity and begin to metabolize and reproduce.

These abilities make some viable viruses candidates for space travel.

Gustavo Caetano-Anolles, an Institute for Genomic Biology professor at the University of Illinois, published a study entitled Giant Viruses Coexisted With the Cellular Ancestors and Represent a Distinct Supergroup Along With Superkingdoms Archaea, Bacteria, and Eukarya in BMC Evolutionary Biology. [64]

With his innovative study, Caetano-Anolis managed to stand near the base of the tree of life and shake up the entire understanding of life itself. He applied studies of past events or "folds" in the three-dimensional structure of proteins of over one thousand diverse types of bacteria, Archaea, and viruses, including the bizarre giant viruses, to study genomes. Genetic sequences can change rapidly due to their instability, and therefore folds are much more like reliable fossils or bones to study.

Caetano-Anolis' groups made a considerably basic biological assumption that the structures which show higher frequency, and in more groups, hence with more diversity, were the older. Many researchers are looking closely at the giant viruses because they seem ubiquitous, implying that they are ancient, have large genomes, and may have been key to the development of life on Earth. Giant virus machinery is remarkably similar to Human cellular machinery and that of most other lifeforms, although the viruses don't use it anymore.

The takeaway findings:

001 There should be four main branches or supergroups of life instead of three, with the giant viruses earning a new and earliest branch.

002 Most ancient protein folds, including mitosing cells, were also found in giant viruses, suggesting they were the first to appear.

003 Giant viruses took a highly significant role in spreading valuable information.

004 Giant viruses evolved from complex to less complex, resulting in smaller genomes with time but more like their original ancestors than their smaller counterparts.

005 The shared protein structures between viruses and other organisms have a very wide distribution. Each of these structures represents evolutionary innovation—thanks to the viruses.

006 Caetano-Anolle's study confirms other research recognizing the importance of viruses, especially the giant viruses, in the "fabric of life."

SMARTER GREAT APES FASTER

Some speculate that alien species, in addition to seeding our DNA in the first place, also tweaked primate evolution later on towards the emergence of Homo sapiens.

If this is true, there should be genetic, anthropological, and archeological evidence. Emily Singer, in Quanta Magazine, wrote an article entitled A Missing Genetic Link in Human Evolution. According to Evan Eichler , a geneticist at Seattle's University of Washington, anywhere from 12 million to 8 million years ago, the common ancestor of the great apes demonstrated remarkable genetic change which occurred in no other family. Some tiny DNA segments, after replication, grabbed some traveling companions and dispersed themselves in a most unusual pattern across the chromosomes. Only the great apes, the chimps, the bonobos, the gorillas, and let's not forget the direct ancestors of humans, benefited from these gene patterns called duplicons. DNA sequencing that repeatedly appears on a specific chromosome defines duplicon. Eichler claims duplicons serve as an efficient mechanism for the creation of new genes.

The new genes assumed important roles in the brain, including aiding and abetting the growth of new cells and contributing to overall brain development and size. One such gene is SRGAP2.

"This mechanism appears to be seminal in our evolution," said Philip Hastings, a geneticist at Houston's Baylor College of Medicine. He emphasizes the descriptive term "seminal" to the duplicon phenomena. This amazing and unique mechanism results in the marked structural change of chromosomes. It is the ultimate source of our intelligence.

Change, of course, did not occur overnight, but it proved itself quite fast in evolutionary measures. Homo habilis evolved from Australopithecus just two to three million years ago, pushing our hominid brains on a path to doubling in size.

Duplicons, because of their complexity, defy testing by standard genetic methods. New ways of study will have to be perfected.

If aliens somehow tweaked early hominids to become smarter, the duplicon strategy of dispersing DNA across the chromosomes would have to be one candidate. This hypothesis is testable. Such change may have tell-tale signs, just as the rest of our DNA gives us signatures and codes to tell us it was, indeed, of alien origins.

Right now, we can only speculate and wait for new testing methods. Was it a coincidence, in other words, a result of random chance, that hominids experienced a mutation that created a unique DNA dispersion system in chromosomes to double their brain size and accelerate their intellectual, emotional, and linguistic development in a short evolutionary time?

Was it a random chance that an ingenious and powerful system arose that no other family on Earth possessed, or would possess, to permit rapid adaptation? Was it the work of little green or grey "eternal" geneticists in a lab on a floating spaceship, bent on seeding beings capable of joining their advanced civilization one day?

MITOCHONDRIA HAVE A STORY TO TELL

The importance of maternal mitochondria DNA inheritance in human evolution and migration studies stands as well-established. Thanks to its maternal transmission, similarities inherited in generations of offspring can be traced to a single line of ancestors. According to research, modern mitochondrial genome fragments can be traced back to a single female ancestor living in a small population in Africa between 150,000 and 200,000 years ago, the time of Homo heidelbergensis, the main candidate for the direct ancestor of Homo sapiens. Scientists generally hypothesize that "Eve" lived among other women in a small population, which increases chances for mutation through genetic drift. Her mitochondria mutated and became an integral part of a new and improved model of early human--Homo sapiens-- through her offspring.

The Homo sapiens package showed radical differences with a 1,500cc brain, larger than modern humans by 150cc, and a unique, elongated shape. The Homo sapiens skull features a short base and its greatest breadth at the top. [65] This unique skull, and the brain inside it, differentiated Homo sapiens from all other previous species. It eliminated constrictive areas directly behind the eyes, or "post-orbital."

Scientists attributed these changes to genetic drift mutation arising from a small population. But all populations were small groups of hunters and gatherers at that time. Might there be another explanation for how such a radical change in the skull and brain, coupled with a more efficient form of mitochondria, arose?

How about an alien intervention? Suppose aliens genetically engineered DNA, as the scientific evidence suggests. Might they also have stopped by their "seeded projects" to monitor them and perhaps alter them towards what they desired, mostly likely a life-infused planet capable of producing a workforce that could not only build infrastructure but could join genetically as hybrids at some time in the future to extend an empire?

Perhaps aliens intervened about eight million years ago with the Duplicons to boost brain size in the Great Apes, again at about 200,000 years ago, to create a radically different skull, brain, and mitochondria, and again at about 40,000 years ago to set Homo sapiens on a path towards the larger groups and civilizations tied to agriculture starting about 10,000 years ago? So what we have done is look through our evolutionary history for a marked change in a short period of time that might represent some form of genetic enhancement by aliens.

BIOLOGICAL SETI

Vladimir shCherbak and Maxim Makukov published an article in 2017 in the International Journal of Astrobiology concerning "Biological SETI."

Given that some scientific literature proposes that life on Earth descended from seeding by an ancient extraterrestrial civilization, a testable hypothesis arises. Microbial seeds sent into space could carry a signature that could endure for extended periods. Furthermore, this signature would be encoded in genetic material since it allows long durability with integrity. It would be the component with the smallest chance for change. It would be the best way to encode messages and signatures found in the genetic codes of all living organisms.

The mainstream scientific community clamored that this proposal is pure guesswork and cannot be confirmed in a meaningful manner because encoding and decoding a signature within the genetic code is not well-defined. Thus, any retrieval attempt, they claim, is doomed to speculation.

In their article, Vladimir shCherbak and Maxim Makukov reviewed the seeded-Earth hypothesis considering recent updates and addressed the possible motivation for encoding signatures. Then they demonstrated that 'biological SETI' involves less robust assumptions than traditional SETI and allows for a well-defined framework of appropriate methods.

Vladimir shCherbak and Maxim Makukov assessed the biological SETI proposal through applications of evolutionary and molecular biology. They formalized the process, adopting SETI's standard guidelines, which stated that encoding/decoding should be derived from first principles and free of conventions. In this fashion, shCherbak and Makukov created a retrieval strategy that proved to be universal. They applied the new strategy to the established genetic code, revealing "a non-trivial precision structure of interlocked logical and numerical attributes of systematic character." Their paper shows how they assessed the result by performing the comparison, statistical, semiotic, and interdependence analyses.

No causal connection of the result to evolutionary models of the genetic code was determined by statistical analysis. Overinterpretation was precluded by interdependence analysis. Comparison analysis demonstrated that known genetic code variations do not have any precision-logic structures, which agrees with these differences occurring post-seeding (i.e., post-LUCA) as evolutionary deviations from the established code.

Signs and symbols (semiotic) analysis revealed that the findings followed the initial assumption. They are logical from the SETI perspective because they "ultimately maintain some of the most universal culture codes." [66]

In 2008, part of Vladimir shCherbak's work on the mathematical nature of DNA appeared in *The Codes of Life* on pages 153-185 as *The Arithmetical Origin of the Genetic Code.* [67]

DNA, according to shCherbak, encodes a privileged numerical system. The genetic code, he believes, is a syntactic arithmetic structure that probability dictates is of artificial origins—it did not arise through natural selection.

The genetic code used for its arithmetic is the decimal place-value numerical system that included a conception for zero. The zero sign governs the genetic code as a necessary part of the decimal system and an acting arithmetical symbol.

Therefore, the genetic code displays an artificiality because of the non-material abstractions, the concept of zero, a syntax based on decimals, and unique summations.

These findings suggest that the genetic code did not arise from chemical evolution on Earth but instead represented pure information derived off-planet.

DNA MODIFIED OR ENGINEERED FROM SCRATCH?

Vladimir shCherbak's 2008 paper claimed that our genetic code and hence DNA arose from pure mathematical information. Did the extraterrestrial geneticists themselves arise from a form of evolved DNA, master genetics and other sciences, and then enhance their DNA, or create one or more different versions for colonizing planets and stamp them with their coded signatures?

Or did the talented alien geneticists arise from a different system altogether, and the genetic code they sent out into space came from a total in-house fabrication?

We need genomic samples from a multitude of planetary lifeforms to test any hypothesis.

PREVIOUS ALIEN VISITATION

The lack of a complete theory of quantum gravity, not to mention the daunting engineering challenges, lack of appropriate materials and tools, would make it impossible for early technicians and engineers armed with slow and clumsy computers powered by floppy disks to make craft capable of UAP quantum gravity and entanglement distortion fields and propulsion along new geodesics. Combining these facts and logical assumptions establishes with near certainty that spacecraft technology is, indeed, extraterrestrial in origin.

That is not to say that humans could not be flying, some of them as pilots or drone pilots now. In The Day After Roswell by Philip Corso, the good Colonel said the United States was bent on reverse-engineering the Roswell craft to make a fleet of its own. They may have been successful many decades after the crash.

No matter the piloting method, the technology of some UAP makes alien visitation of Earth indisputable.

We will never forget the famous “weather balloon” of Roswell. My father heard the announcement of a recovered flying saucer on the radio, which was retracted the next day.

Can we trust the government to tell us what we want to know?

Some of the most convincing evidence the government released has to do with the bizarre materials in their possession. They allowed a contractor civilian, Eric Davis, to make statements about these materials being not created on Earth, a way to inform the public that UFOs are real, that aliens have visited Earth, and save face at the same time.

Synthetic DNA, in 2019, formed a larger version of *E. Coli*. Although in its nascent stages here, DNA engineering shows great promise in various applications, including medicine. [68]

It is not too farfetched that Human DNA may be hiding genes of more than one ancient alien civilization. In that case, we are hybrids of hybrids.

CRICK'S CONCLUSION

From his knowledge about DNA and early conditions on Earth, Francis Crick thought that DNA did not originate here and published his opinions in the peer-reviewed journal *Icarus* and the book *Life Itself*. The results of natural selection, although overall impressive, often arose from a clumsy hodgepodge of what came before, often filled with redundancy, inefficiency, and incapacitating short-circuits.[69]

DNA demonstrated such elegance, efficiency in its functions, such "perfection," and the ability to carry the same message for billions of years that Crick doubted it could have arisen from a few simple chemical complexes and gases through natural selection. It would have had to do so in less than three hundred million years—much too short of a time.

No, DNA must have been artificial. In other words, DNA was the result of a special mathematical and engineering project to create and spread life.

He presented sound arguments, but he had no proof at the time of his writings.

Vladimir I. shCherbak and Maxim A. Makukov came along in 2013 to offer proof that DNA represents elegant mathematical coding, activation keys, pictorial communications as some of many signatures that DNA, like Crick suspected, arose billions of years ago through intelligent design and not from Earth.

New work from new methods in evolutionary and computational biology, genetics, and virology points to a Last Universal Common Ancestor (LUCA,) an ancestor of the giant viruses, consistent with Crick's theory of a panspermist seeding. Many giant viruses contain many genes found nowhere else on Earth but also genes found in all life.

Crick concerned himself with Enrico Fermi's question: if advanced civilizations exist, why don't we encounter them?

Astrophysicist Eric Davis of Bigelow Aerospace claims that the United States government possesses bizarre materials from alien spaceships with remarkable properties that humans are incapable of making at this time. The government is studying people who claim to have experienced encounters with aliens or their ships. The government also released several films of UFOs demonstrating amazing aerial maneuvers that no known Earthly aircraft can emulate.

Although we still have no absolute proof from the government that aliens have visited Earth, we are several steps closer.

In any case, coupled with a five-thousand-year history of sightings of strange flying objects and discussions of possible "aliens," some of it even set in the stone beginning with Sumeria, the recent developments in the last two decades suggest that it is more likely that they have visited rather than not.

Fully a two-thirds majority of Americans, as of 2021, believe aliens exist. [70]

We might determine which planets are likely to be hospitable to the start of life. Then, if possible, rockets capable of reliable long-distance travel and fast ones seem reasonable to send out with colonizing spores. Sending instruments to report back to us first also seems reasonable, but why not also send microbes?

What about cosmological ethics? What if life existed on a target planet already? Do we have the right to disturb it and compete with it? Conquer it with a parasitic giant virus?

Crick concludes the epilogue by stating that we should be mindful of the possible negative consequences of Directed Panspermia. "Contamination" and "infection" reveal themselves as unpleasant words and concepts when dealing with other planets and lifeforms.

SECTION: ROSWELL

Based upon the crash of one or more alien craft near Roswell, Nevada, Lieutenant Colonel Corso shared with the world the nature of the craft, its methods of locomotion, the nature of its cloned pilots and crew, some of its material structure, and multiple forms of advanced technology that would transform industries and economies.

Corso's The Day After Roswell lies at the heart of the modern history of the alien presence on Earth. The history of UFOs and aliens becomes the dominant story of our planet, affecting all of politics, social and intellectual development, and economies.

In a lightning storm in the middle of the night, the crash at the remote town of Roswell, New Mexico, stands as arguably the most important single event in human history.

The technology derived from the crashed spaceship and the nature of the aliens as clones, penetrated by that technology, would lead to a worldwide tech revolution.

The Cold War with the Soviets would lead to the Strategic Defense Initiative, which would help catalyze the end of the Soviet Union.

The Roswell crash earned global publicity for the topic of aliens, and now, with the new evidence of UAP, it would answer one of humankind's ultimate questions—we are not alone in the universe.

ROSWELL'S HERO

Philip Corso matured into a decorated and trustworthy military man. He advanced to be one of the top intelligence officers in the Army in research and design of "foreign technology" at the Pentagon. This broad classification included techno-artifacts from UFOs. [71] He boasts nineteen medals in twenty-one years of service in the Army. According to publisher Simon and Schuster, a person of exceptional prominence in the national archives in St. Louis, Missouri, Corso served four years on President Eisenhower's National Security Council. [72]

Colonel Corso felt obligated in his eighties to pass on to his children, grandchildren, and the rest of the world, some of the details of his most unusual, planetary significant, and historic career. It took him five years of writing to finish his book. He said in his memoirs that he hoped his book would help prepare Humanity for any hostile alien presence.

Because of the secrecy insisted upon by the military for over nine decades regarding UFOs, Colonel Corso, in a videotaped interview on February 1, 2008, said that, if he did not share it, the story would die with him. [73] In the same video, his son said his father shared only part of what he knew in his bestselling book, The Day After Roswell. The rest remained buried under a top-secret classification.

Corso explains what the government would not do on that dark desert night on July 3, 1947, during a fierce lightning storm near Roswell, New Mexico. Either from the lightning, or the effects of a multitude of radar beams tracking it, or both, an alien spacecraft crashed, spewed out at least five aliens, two of them alive, and all of whom had been genetically engineered for space travel.

The amazing craft was comprised of such advanced materials that the vast majority of it remained intact. One large crack in it allowed the aliens to either exit on their own or be pulled out.

Corso's story becomes remarkable, enlightening, and unique—and quickly. He first describes what happened at Roswell in July of 1947. Next, he focuses on his role in the commercial development of technology gleaned from the ship and the aliens themselves, all against a backdrop of secrecy to protect the technology from the Soviets. Communists had gained a solid foothold in the CIA after World War II. Corso's book morphs into an insider's view of United States history from World War II through the 1980s. What he reveals is astonishingly relevant for today's world and our survival as a free nation. Corso shows the reader how the government works and how we fought a war against aliens as we fought a cold war against the Soviets.

The way the military handled the commercialization and full use of the Earth-changing technology so as not to draw attention to its origins while at the same time censoring, bribing, and intimidating witnesses, denying the existence of UFOs altogether, and propagating disinformation is not pretty, and was a bit sloppy. Nevertheless, it proved to become the biggest successful coverup in the history of the United States.

Corso, respected and well-connected in the Army, stood right in the middle of it all, mouth shut, ears wide open, quietly doing his job despite tails from the CIA.

A series of unusual and unlikely circumstances just happened to place Corso in charge of the night watch on Fort Riley, Kansas. He could go wherever he deemed fit as he checked each station. Corso happened to have this assignment just a few days after the Roswell crash. The dead aliens were en route across the country by truck. The truck stopped at Fort Riley.

As Corso made his security rounds, one of his bowling friends had been on watch over the alien cargo and motioned for Corso to take a quick look inside some packing crates in an off-limits facility. What Corso saw created an unforgettable, unpleasant memory for life. It was an exceedingly small and very dead alien with an oversized head and large eyes. Corso said nothing and told no one, but fourteen years later, he found himself entrusted with a file cabinet full of documents and materials about Roswell from the Chief of Research and Design, Lieutenant General Arthur Trudeau [74] . Corso had worked directly with President Eisenhower as a National Security Council member, handing him special intelligence reports and waiting for his reply. When Eisenhower retired, he passed Corso on to General Trudeau. Trudeau became Corso's next boss in Army intelligence. Later Corso would view the results of the autopsies on the aliens. When he retired, Trudeau served as President of Gulf Labs for the Gulf Oil Corporation until 1968 in Pittsburgh . Trudeau's next position would be a special adviser to the chair of Rockwell International. Trudeau became a legend in the Korean War for trading his helmet for one of a lower rank so he could lead a charge, under heavy fire, up Pork Chop Hill to bring some of his men back to safety.

Corso's job: get the ball rolling for gleaning commercial uses of the remaining alien technology and get it into the hands of the right contractors—but secretly. The Air Force formed immediately after Roswell had grabbed the craft. Bell Labs procured some of the futuristic electronics. On July 27, 1947, the National Security Act created the US Air Force and CIA as independent agencies.

Corso consulted with several top German scientists who had decided to work with the United States after the war. Corso credits the Roswell craft from transistors to integrated circuit chips, from chips to smaller and faster computers, from computers to the internet. Kevlar, lasers, the stealth bomber, fiber optics, night vision goggles, and particle beam weapons also arose from what the craft had to offer. In other words, the long-term development of the technology from the craft led to the electronics, computer, and Silicon Valley revolutions, revolutionary aircraft, facilitated space travel, and sophisticated weaponry.

Corso reveals why the military immediately labeled the alien crafts as a threat. The aliens spent considerable time buzzing top-secret military installations, including those harboring nuclear weapons, for reconnaissance. They had no hesitation in extracting a uterus or other organ from more than a few unfortunate cows, with laser scalpels, like the one found on the ship, or abducting Humans for testing and experimentation. Kidnapping people against their will, and running medical tests and experiments which terrorize them, are not behaviors of friendly, benevolent aliens.

Our astronauts and spacecraft found themselves painted with bullseyes. The aliens jammed critical communications and scared the hell out of our astronauts with close flybys. They didn't want us on the moon at all. Corso not only claims that they forced us to end the Apollo program but that they had already built a base on the moon. Our government feared they evaluated us for a takeover. They didn't share their technology with us. We had to take some of it from them. We were forced to use it against them, Corso claims, for our survival. We used accelerated particle beam weapons, lasers, and stealth aircraft forms to hold them at bay.

The suspected genetic engineering of the alien pilots and spies relates to Section: DNA in this book. Francis Crick claimed that Earthly DNA arose from elsewhere, that it had been genetically engineered by advanced aliens billions of years ago.

Corso's claims suggest that the Roswell aliens demonstrated an elevated level of expertise in genetic engineering. One theory: ancient aliens seeded planets for life with engineered DNA for harvesting later as infrastructure for colonies, with hybridization with any local intelligent species, not a bad Darwinian strategy for spreading one's genes and culture across the universe. Of course, we don't know if the Roswell craft inhabitants represented descendants, or engineered projects, of the same aliens suspected of seeding life on Earth. If they were, we would have a much better understanding of our beginnings—and futures.

One thing that we would like to know. Were the aliens based on some form of DNA?

Corso would wait for him to pass away before publishing out of his respect for Trudeau and Trudeau's family.

PHILIP CORSO'S MONUMENTAL TASK

Colonel Corso worked for the Army as an evaluator and researcher, specializing in "foreign" weapons systems. He examined technical reports and consulted with a variety of engineers and scientists. He reported to Lieutenant General

Arthur Trudeau, the chief of army intelligence, research, and design, who oversaw the activities of over 3,000 enlisted personnel.

Under the guise of routine work, Corso led a double and surprising life, reviewing highly classified information and acting upon it as directed by Trudeau. He found himself pursued by the CIA and KGB and had to know the games of spies to do his job.

Corso knew of the KGB's penetration of the CIA and other government departments, influencing policymaking.

Because of his experience, dedication to the military and the United States, history with President Eisenhower, and intelligence expertise, Lt. Gen. Trudeau entrusted Corso with the remaining Roswell artifacts and documents in 1961. Roswell quickly had morphed into the Army's most important secret after the crash in 1947. Corso reported to Trudeau and not Admiral Roscoe Hillenkoetter, the Army's director of central intelligence, who would become the first director of the CIA.

As part of his job, Corso read the reports on Roswell from a variety of military sources. The bottom line: the crash of an alien craft at Roswell, with aliens present, stood on record as an indisputable reality.

Lt. Gen. Trudeau laid out Corso's assignment: utilize the Army's weapons research and development program to sift through the remaining technology found in the craft and from the nature of the aliens, and hand it over to the appropriate long-standing and trusted private contractors for development, including for military applications. Adding the sacred technology to ongoing contracts helped mask what was going on instead of creating new ones.

Not only did the military label the aliens as an existential threat to Humanity, but it also feared the aliens had interfered in World War II. Corso claims the military and the United States government it represented considered themselves at war with Soviet communists and aliens.

Although Corso did not speak of it in his book, we always have to think of who benefits the most from organized wars, and especially those started with false flags. The United States was financially devasted by WWII and allowed itself to be drawn into agreements for occupation by aliens while the communists started and dominated the CIA. One possibility is that the Grays worked with the communists to create and profit from WWII.

The processing of alien technology led to the alien defeat, albeit a temporary one, by military efforts presumed to be in concert with the Russians, which fostered the end of the Cold War.

Corso credits a handful of officers in military intelligence for altering the course of history for Humanity. He stood proud to be among them, just "a little man from a small town" in Pennsylvania, who ended up with one of the biggest jobs in the world.

THE ROSWELL DESERT AND CRASH

Pitch black skies penetrated by fierce seasonal bolts of lightning and rainstorms are the norm rather than the exception on the lonely drive to Roswell, New Mexico.

These desert lightning storms played a huge role in what happened on Earth from July 1947 to today.

Corso researched the many versions of what happened with the major crash. Here is a summary of his final evaluation, which represents a consensus of the "facts" to the best of his knowledge.

- July 1, 1947: strange radar blips, representing craft flying at high speeds and making unbelievable moves at the Army's 509 th airfield, Alamagordo, and White Sands. Pursuit by our aircraft proved futile.
- Local civilians experience UFO sightings.
- World War II operatives and intelligence personnel (CIG) descend on Roswell to witness and plan accordingly.
- The odd blips continued for days until the screens represented an explosion of one of the aircraft.
- An extensive team found themselves immediately en route to the desert to retrieve anything and everything from the wreckage.
- Campers, ranchers, and residents had reported, that same night, a mid-air explosion and a brightly lit object descending towards the Earth near the town of Corona.
- To the military's dismay, local police and the fire department are deployed to the site of the crash just as quickly.
- The crash showed plenty of small bits of debris, but the craft itself laid intact with but one large visible crack.
- The crash occurred the night of July 3, 1947.
- The crafted seemed small and appeared like an airplane wing.

- Three small humanoids, about 4.5 feet tall, dark grayish, laid on the ground, dead.
- Another lay against a small hill in the desert sand, and another near the craft.
- The Army's Steve Arnold peered inside the craft and later said it looked like daylight inside.
- The alien at the sandhill partially recovered and tried to make a run for it but was stopped. Corso originally said it was shot but retracted this after publication in an interview with Art Bell.
- The Army's crew had orders to secure the area, discourage the presence of police and fire-fighting personnel and all other civilians, and pack up the aliens and their craft and get them to the base, and pick up all of the debris on their hands and knees if they had too.
- The craft, surprisingly light, found itself clumsily placed by crane onto a flatbed and covered with a tarp.
- The aliens had oversized heads and eyes, tiny mouths, ears, and nostrils, appeared gray or brownish-gray, and had no hair,
- Fireman Dwyer allegedly "stole" a piece of alien fabric that could remember its shape if folded and which he would show to his daughter, who would describe it to the public on television many years later.
- The civilians who visited the site were instructed not to discuss the incident or anything they saw with anybody.
- 509 th commander Bull Blanchard authorized the dissemination of a story featuring a crashed "flying saucer" that quickly went worldwide but would be retracted by the next day and replaced by the famous "weather balloon" fabrication which a reluctant Major

Jesse Marcel would tell and who would retract his story towards the end of his life. He would confirm that it was, indeed, an alien spacecraft that crashed.

- His son would eventually share how his father showed him some of the debris, including a metal rod with purplish hieroglyphic-type symbols engraved upon it.
- Threats, including murder, bribes, and intimidation of all witnesses, followed for months until everybody had agreed to shut up about what happened that night.
- Some debris and the craft were flown to the Army's Wright Airfield in Ohio, some to Fort Bliss, Texas, and the rest convoyed by truck to Fort Riley in Kansas.

Author's Note: later in Corso's book, the question would arise if the captured aliens reported to a mothership because of the small craft's absence of life support systems. Suppose a mothership, indeed, proved to be present somewhere. Why would it not make an effort to recover its crashed spy craft immediately, knowing that the Humans could eventually duplicate some of the technology that could be used against alien fleets? Why was there no confrontation or battle that same night in Roswell over the crashed UFO? Was there no mothership? Or was it too late to act without drawing even more attention to the aliens' presence?

SEEING A ROSWELL ALIEN BODY

After a job well done, including a meeting with the Pope to discuss civil engineering for rebuilding Rome, Corso returned to the United States to attend Military Intelligence School at Fort Riley, Kansas. Corso was no novice in intelligence skills. He had training from MI10 from Great Britain and had worked to undermine Soviet intelligence in Rome.

Corso happened to be excellent at bowling, which led to something that occurred on July 6, 1947, that he would never have even dreamed about. He had made friends on the Army's bowling team, including a man with the nickname of "Brownie."

Corso happened to be assigned to be in charge of the base watch on the night of July 6, 1947. Brownie, Corso noticed on the roster, guarded the veterinary building at Fort Riley that was marked "No Access" for that night—it was officially off-limits to everyone, except the commanding officer of the watch—Colonel Corso.

Earlier that day, cargo en route to Wright Airfield had been unloaded and stored, temporarily, at Fort Riley. It turns out that it had originated from the 509 th near Roswell after processing at Fort Bliss. Corso didn't know it at that time, but cargo always went from Wright to Fort Bliss, not the other way around.

On his rounds, Corso had to stop at all checkpoints, and he stopped at Brownie's, but to his surprise, Brownie was not to be seen, an obvious breach of duty. Brownie soon spoke from the darkness and pleaded with Corso to come inside the building and view the newly arrived cargo.

Soon enough, Brownie revealed himself to Corso from inside the veterinary building, another breach of duty since the interior was officially off-limits for Brownie. Brownie guided Corso to what he thought he should see and took a position guarding the entrance.

Corso started to open the crate that looked like it had been opened recently. Inside, immersed in a blue liquid, floated a deceased alien. The crate he opened contained its coffin. He read the paperwork enclosed inside to confirm its place of origin and what they thought it was—something from a downed aircraft. It was to be delivered to Walter Reed Hospital for a presumed autopsy.

Corso's description of the alien:

- The body looked shiny and soft, like the belly of a fish. It was pale grey, but Corso didn't know if that was its skin color or skin-tight protective clothing.
- The humanoid body was about four feet long. It had two arms and hands with no thumbs—only four fingers.
- It had two thin legs and skinny feet.
- Corso saw no eyeballs, only deep-set, oversized, and elongated eye-sockets. He saw no irises or pupils.
- Its oversized head had two nostrils for a nose and small slit-like openings where the mouth and ears should have been.

Corso left, shocked, knowing it had to be an alien, and ordered Brownie never to discuss anything he had seen with anybody.

THE ROSWELL ARTIFACTS

Corso reported in 1961 to General Trudeau for his new position in Pentagon intelligence after commanding an Army anti-aircraft missile contingent in Germany. Having met him at the White House, Corso had known General Trudeau while serving on the National Security Council.

He knew Trudeau from working for President Eisenhower,

This was when the KGB had great success in influencing American policy through infiltrating various government agencies. Their actions weakened us and strengthened the Communists. The CIA became riddled with KGB moles. If we wanted the KGB to know something, whether true or false, we simply told the CIA. Corso wrote about a secret government within the government, what we call the "Deep State" today.

Corso believes we lost the Korean War because of leaks to Russian advisors to the North Koreans and China.

Trudeau sat Corso down and told him he would be his special assistant in research and design. This proved to be the department for the highly classified scientific documents and blueprints for weapons that would fuel lucrative defense contracts. Corso's job would be to keep the remaining Roswell technology secret as it was to be developed commercially to be fed back to the military and key industries.

Trudeau put Corso in an office directly below him, one floor down, near a stairwell close to both men's offices. Corso could "sneak" into Trudeau's office and back out again without drawing much attention. Corso would be in charge of Foreign Technology research for weapons, as he found out that the new file cabinet sitting in his office housed Roswell's "foreign technology." Trudeau made it clear how valuable it was and that the CIA, KGB, Navy, and Air Force all wanted desperately what sat quietly and secretly in Corso's new office.

Corso would mix with specialists and defense contractors to figure out what Roswell could be used to help advance the United States.

One of the first pieces from Roswell that caught Corso's eye revealed single filaments that were clear, flexible, and twisted together as if they were cables to be fed into some type of junction. They could carry light along their length. (They were something resembling today's fiber optics.)

Corso also examined two-inch wafers with small etchings for circuits that would later catalyze the modern integrated circuit industry. His placement of the wafers would result in the computer and cell phone revolution and the supercomputer thirty years later. [76]

Corso paid considerable attention to two "eyepieces" that the aliens used to their great advantage against us, and that would lead to night vision goggles. They were like third eyelids. With them in the proper orientation, one could not only see outlines of objects in the dark but literally "see" outlines through walls.

A piece of "cloth with metallic fibers," and quite strong, resumed its original shape no matter how much it was crumpled or folded.

A "surgical cutting device or weapon" turned out to be a hand-held laser. It was to Corso a "light beam torch" and became handy in extracting the organs from cows, like the uterus, eyes, tongue, and patches of skin, without leaving any sign of trauma.

Corso examined a headband that was believed to help the aliens fly their spaceship with their minds. The pilot and the ship became "one" electronically.

He learned from his consultants that one gadget might have been a potential brain implant for fusing with artificial intelligence.

Corso's assignment: keep everything from Roswell hidden from the CIA, KGB, and the rest of the Deep State, and get anything with military and commercial applications to the appropriate contractors under existing contracts so they could patent and manufacture it.

The haunting memory of the alien Corso had seen in 1947, some fourteen years earlier, had not left him and now manifested itself into a different form.

FOREIGN TECHNOLOGY DESK

Corso likens the Pentagon to any large organization with too many departments with too many agendas and goals, all in competition with each other for limited resources.

Corso found himself smack in the middle of it all as a weapons development person during Kennedy's and Roosevelt's Cold War with Russia.

While the public remained shielded from the truth about Roswell, the military brass and our enemies knew quite a bit. Each major division had some pieces already, and the "war" now was about laying hands on more technology, what they called the "alien harvest," or "the cargo," or "the package," or "junk pile," or what Corso referred to as "the nut file," that could shift power, bring in massive funding, and quickly.

Corso describes the simple techniques employed by the military to continue to cover up Roswell for decades. Bribery, physical intimidation, murder threats—about anything was justified to keep the technology and alien story hidden.

President Truman gave General Twining the job of handling Roswell. Twining evaluated all evidence and data and recommended continuous study of the entire matter that he determined as real and a threat to Humanity.

Corso then quotes the names of the people who would form the "Majestic 12" to oversee the handling of the Roswell incident, the UFO phenomenon, and aliens, in general.

THE COVERUP

Russia's Stalin proved to be a big believer in UFOs after seeing the initial Roswell headlines of the Americans recovering a flying saucer, which the KGB confirmed. [77] He knew the implications. America would become the undisputed superpower in the world if it could harness the powers of alien technology.

Stalin only hoped to learn about the technology that arose from moles inside the US government and US bases. The US government knew this right away and began a counterintelligence campaign immediately. Anybody who asked questions about Roswell would be questioned themselves. To successfully hide Roswell from the Russians, Roswell had to be hidden from the US government itself. Under no circumstances would the public become informed of the truth.

The Majestic 12 produced Project Blue Book and other deceptive programs, which would permit the government to continue to gather information on UFO sightings, materials, and abductions, some of it quite real, while at the same time publicly dismissing all information as bogus. They also would form a type of Deep State, a secret government within the government, knowing what was going on while shielding all others from the truth.

THE STRATEGY

Corso shares how secrets in the government remain secrets. Cooperating branches each house only part of the secrets, buried under layers of classification and bureaucracy. New branches arise continuously to snatch secrets away from earlier branches that may be vulnerable or compromised.

If a secret became found out, a partial disclosure might ensue, but the bulk of the secret moved to a new location.

The truth claims Corso, rather than being "hid," wrapped itself in such camouflage that it would not be recognized when found.

Project Blue Book is a perfect example. The public believed they had a viable and fair way to report sightings while the government usurped the legitimate ones as it dismissed them all as illegitimate and encouraged ridicule. Blue Book sightings took attention away from Roswell—exactly what the government wanted.

Alien technology themes underlay the Space Defense Initiative, primarily hatched to defend against aliens rather than Russia, which, of course, even with allotments in the trillions of dollars, the public never figured out.

Washington, DC, found itself the setting for a flyby by UFOs in 1952. Photographs from multiple perspectives, hundreds of witnesses, and numerous radar reports confirmed it. Still, the government denied the entire event and pushed Hollywood to make movies like The Man from Planet X instead.

Corso's term for this response: "Camouflage through limited disclosure" meant to keep the public occupied with bits and pieces of a real story, or a total fabrication, to distract them from discovering the truth.

General Trudeau pressed Corso for information about the Majestic 12 that he might have learned while working for President Eisenhower in the National Security Council.

Corso believes the group became too thin and had to farm some of its duties to recognized departments. He also thought a power struggle began to permeate the group. The result:

many great departments handled bits and pieces without any true oversight by a commanding power. Individuals made decisions about what should be done, like General Trudeau, without answering anyone.

To make the coverup even more complicated, J. Edgar Hoover began his investigation of Roswell.

Corso also claims that the Roswell artifacts remained inaccessible to the CIA and most governments in the research and design department for years and years.

Corso also figured out that the KGB, like our CIA, kept information away from the larger "official" Russian government and its military. They acted to preserve and empower themselves as priorities. FOR THEIR SURVIVAL, the KGB and the CIA had more respect for each other than their respective governments, almost like part of the same team. The exchange of information between the CIA and KGB, according to Corso, prevented a nuclear war with Russia.

The CIA "tailed" Corso during his four years under Eisenhower and, in 1961, resumed a tail when he went to work for General Trudeau. He thought Frank Wiesner, the director of covert operations, acted for the Russians and bore responsibility for the tail. Showing his disdain for being followed, Corso stormed into Wiesner's office one day carrying a .45 automatic, laid it on Wesner's desk, and told him the next tail might have both eyes shot out.

Corso's and Trudeau's plan to handle the "nut file" revolved around a business as usual philosophy to keep everything looking ordinary and well-under the CIA and KGB radar.

Corso had a handful of German scientists he already trusted for consultants, like Wernher von Braun and Hermann Oberth. Corso would submit Roswell technology to long-standing contractors already on the payroll. The contractors or their staff would file for the patents, further masking the true origin of the technology.

Proceeding to create reports for presentation to Trudeau, Corso summarized the details of the autopsies from Walter Reed Hospital on the aliens, each of which they called an Extraterrestrial Biological Entity or "EBE":

- Clones
- About four feet tall.
- Larger lungs and heart than Humans.
- No vocal cords
- Thinner but sturdier bones, with substantial tensile strength.
- Outer protective "skin," part synthetic, offered protection for vital organs against strong electromagnetic waves, cosmic rays, and high gravitational forces. Skin atoms "aligned" with spacesuit atoms could even protect from cosmic radiation. Also aligned with the craft, which itself served as a biological entity.
- Larger brain in proportion to the body.
- Technology implanted in the brains to enhance telepathy.
- "Third eyelid" that permitted night vision.
- Striking and disturbing similarities to Humans.
- Likely genetically engineered for high-speed space treks of long duration.
- Conjecture came from Hermann Oberth that the vessel could also serve
- as a time machine with his proposal of how the propulsion/repulsion system performed, including diagrams.

- Rather than evolved lifeforms, the EBEs seemed more like humanoid robots that were engineered for space and time travel.

Corso claimed he saw one craft at an Air Force Base in a televised interview but did not go inside. [78] He said he was aware of steady, intelligent signals received from space as verified by the NSA in the same interview.

THE EBE

Corso realized that further study of the EBEs could lead to Humanity traveling more successfully and farther in space. He thought that this should be the "overriding concern" for all of the artifacts. Autopsy and examiner conclusions from Walter Reed:

- The huge capacity of the oversized heart and lungs implied an extremely slow metabolism responsible for a quiet passage of biological time—ideal for long-duration space travel. Their "blood" looked more like thin and milky lymph, and there was not much of it.

- The heart, with less of a workload than a Human heart, partly because of the miniaturized body and partly because of metabolism slow, seemed functional with only internal and diaphragm-like muscles.
- Because of the rapid decomposition of the heart and other organs, they surmised that our atmosphere proved to be toxic to the EBEs.
- Bones were thin, fibrous, flexible, and resilient. They could act like efficient living shock-absorbers.
- Forces that would shatter Human bones would be of no consequence to the EBEs.
- Observers at the crash site said the two EBEs that were alive showed difficulty in breathing.
- Given the circumstances, the examiners and military brass doubted that these particular EBEs could ever exit their ship as they seemed unprepared to do so.
- The EBE mouth looked like a small slit. No apparent food was found onboard. Not big eaters. The EBEs sucked up whatever minimalist nutrients they needed another way.
- The EBEs, too many, seemed, eerily, to be reminiscent of little Humans.
- Although the initial examinations revealed no new basic elements, Corso's team suspected looking for strange combinations of organic compounds.
- The blood "fluid" served as a regulation mechanism for bodily functions, much like glands do for us. The blood and lymph circulatory systems appeared to have been

combined into a glandless, energy-saving, and very efficient system.

- With no waste or digestive systems, the skin or their protective clothing took over those roles.
- The outer one-piece space "spider-web" jumpsuit boasted tremendous flexibility and tensile strength. The fit reflected perfection and tightness that it must have been "spun" on rather than "put" on.
- Fibers in their Spider-Man suits ran lengthwise, shielding the little cosmonauts from harmful rays and the inevitable bombardment of toxic subatomic particles.
- They suspected that the magical suit also protected their vulnerable and fragile, and few, internal organs.
- EBE inner skin featured a fatty tissue layer, very thin and extremely permeable. It constantly exchanged whatever was needed, including food and waste, with the spacesuits. (Authors Note: with such sophisticated genetic engineering, it would be likely that there was little waste at all.)
- Corso considered them androids or robots, neither of which would need food like we would or have waste to dispose of, anyway.
- Walter Reed thought the creatures might return to a mothership frequently enough not to need to store food or waste onboard.
- People at the scene could sense emotions from the EBEs like pain, sadness, or suffering but heard no words in their heads. No sounds at all were heard from them.
- EBE brains had four distinct lobes, and the skulls felt more like spongy cartilage than bone.

Headbands, made of a flexible plastic-like material, functioned with the embedding of sensors and electrical conductors, paying some tribute to the assumed powers of the EBE brain. These headbands served as communication devices and a way to pilot the ship through brainpower.

Army helicopter pilots, decades later, would employ a similar device.

No traditional controls or propulsion systems revealed themselves in the craft, which implied mind-control and electromagnetic waves or proficiency in gravity and spacetime.

Norton Air Force Base in California would secure the craft from Wright Airfield for replication and evaluation for determining defensive strategies.

The craft carried weapons, and the Army was determined to reverse engineer them. In response to the gravity of Roswell and labeling the aliens as an existential threat, the Air Force and CIA would be formed just a couple of months later in September under the National Security Act.

Corso shares with the reader that the reverse engineering process for the craft paraded itself in front of select members of Congress to make an impression on funding. Roswell became the ultimate game centered on planet-changing toys.

Corso's team thought the craft displaced gravity through electromagnetic wave propagation and shifting magnetic poles to create a repulsion force. The craft itself served as a giant capacitor. The modern analysis concludes that the aliens mastered quantum gravity and entanglement to bend spacetime near the ship to travel along desired geodesics.

It required decades and billions to succeed in reverse-engineering the repulsion mechanism finally, and the air force hoped an entire generation of new air and undersea craft would hit the skies and seas.

Gravity, rather than crushing the craft's occupants, folded itself around the perimeter of the electromagnetic wave, bending spacetime. No wonder the crafts could make 90-degree turns at high speeds without living pilots and crew suffering from devastating G-forces. The occupants experienced no acceleration on their bodies.

Corso's team suspected that the creatures' suits, with their lengthwise filament alignment, made the pilot and crew a requisite part of the repulsion system. The pilot and crew integrated themselves into the craft's electronics, and their nervous systems controlled the ship's functions. The pilots became the primary circuit of control.

Corso lays out the tremendous implications of harnessing the full gamut of the Roswell craft's technology and the nature of its pilots and crew. The EBEs dropped some of the keys to advanced technology, space travel, and genetic engineering right at our feet.

All we had to do was pick them up and get to work unlocking the doors to a most intriguing future and Star Trek's final frontier.

THE PROJECT GETS UNDERWAY

Corso depicts an impressed General Trudeau eating up Corso's first official report that had focused on the aliens themselves and the prospects for extended space travel.

Looking so tired that the General thought Corso had stayed up all night writing it and commenting as such made Corso admit that is exactly what happened. Corso had been too excited and too proud of his work to be available for anything so insignificant as one man's sleep.

Trudeau needed more to maintain funding. Corso revealed his plan: work with the scientific experts who not only could be trusted but were free thinkers as well.

- Robert Sarbacher
- John von Neumann
- Hermann Oberth
- Hans Kohler
- Werner von Braun
- Wilbert Smith

Werner von Braun knew about what happened in Roswell from working with General Twining back in the late 1940s. All of the others knew about Roswell from their connections with the Research and Development Board during Eisenhower's administration.

Corso would categorize the contents of his nut file and consult with the scientists who would suggest appropriate applications. Dressed as a civilian, Corso would quietly take the proposals and artifacts to the applicable manufacturers under contract with the government.

Although the team might know what something was and what it was used for, replicating it exactly was another problem altogether.

The engineers at Norton Air Force Base examined the craft. They found that its underside proved to be covered with a very thin amalgam of the purest silver and most refined copper imaginable—perfect for maximum conductivity with little to infinitesimal resistance.

However, with current technology, they could not duplicate it.

Without the magic of a futuristic repulsion mechanism to power the craft, attempts to replicate its performance with other types of engines failed well into the 1960s.

Corso reviews the similar level of conductivity found in the aliens' skin suits which could vector electric fields. They also knew the headbands had to navigate the ship, but exactly how remained unknown.

Corso's experts figured that the answer might lie in how the aliens' brains, skins, and the headband somehow linked to the ship itself as a huge capacitor to tie everything together so that mere thoughts and hand placement on an unearthly control panel could power and pilot the most unusual craft. The ship did have raised deck panels littered with indentations for EBE hands.

The capacitor would continuously recharge itself in flight—no need to stop for gas. Anti-gravity flight had been on the drawing boards even before World War II.

Corso reviews the problems. Humans encounter space travel, like suffering from the G-forces from launches, then the harmful effects of prolonged weightlessness. He recommended early training of astronauts, even school age.

Potential projects summarized by Corso from the artifacts from Roswell:

01 Fiber optics

02 Depleted uranium projectiles

03 Night vision and "see through walls" goggles

04 Electromagnetic and gravity wave propulsion/repulsion systems

05 Supertenacity fibers

06 Particle Beam Weapons

07 Lasers

08 Headband guidance systems

09 Pure metallic alloys. (Titanium research and development accelerated right after the Roswell event. [79] Titanium and nickel comprised the "memory metal" or Nitinol)

10 Food that was irradiated

11 Microminiaturization of logic boards and integrated circuits

12 Ion propulsion drives from portable atomic generators

13 High Altitude Research Project (HARP)

14 A moon base (Project Horizon)

15 Artificial Heart Pump

Some of the corporations they would approach:

- Bell Labs
- Hughes
- General Electric
- Dow
- Monsanto

- IBM

HOSTILE INTENTIONS AND THE OTHER COLD WAR

Corso spent considerable time at Fort Belvoir, located within a thirty-minute drive from the Pentagon. Fort Belvoir, which housed the Army Engineering School, ran some of the top-secret research on UFOs. They stashed movies of retrievals of crashed craft by the military. The Air force even ran an elite unit explicitly with protocols for retrieving alien ships and aliens.

Corso admits that the military denied all extraterrestrial activity to prevent mass panic and temporarily relieve themselves of any responsibility to respond to the real threat while they prepared to fight it. It was easier to deny abductions than to admit that the government could not protect its people from superior aliens.

Our government took advantage of the Russians' and Chinese' historical acts of aggression to garner needed funding to build enough firepower. The Cold War with Russia allowed for a planetary tracking and defense system.

President Reagan made it clear to Mikhail Gorbachev that we could deploy such an effective missile defense system with the Star Wars technology to neutralize an attack from Russia. Reagan argued that Russia and the United States should both point their weapons out towards the aliens. Hence, the Cold War of the 1950s and 1960s would end or be held until the strange situation was resolved. (Author's Note: Reagan, in three speeches, including one to the United Nations, proposed that an attack by aliens would unite the people of the world such that they would quickly end their battles and accept their differences. [80] [81] Pure coincidence? Corso argues that war had already happened, and his version can be characterized as consistent with historical events.

Many people wondered why Russia and the United States needed at least ten times more missiles than necessary to destroy all major cities in both countries. They didn't. They needed them to deter an alien attack. If the aliens attacked and secured territory anyway, we could simply blow them up with the nukes and sacrifice that territory to preserve the remainder of the planet.

Corso then reviews the aliens' hostile acts, including against our space programs:

- Disrupted worldwide communications and defense systems, including flying over and disarming missiles.
- Made communications targets out of the Mercury and Gemini space programs. Knocking out communications endangered the lives of the astronauts.
- Cattle mutilations that suggested precise laser surgical tools and highly skilled surgeons.
- Abductions bore common patterns despite abductees' lack of knowledge about abductions, in general.

The government's conclusion: we were under a probing attack from aliens who tested our abilities to fight back and measured our resolve to do so. Therefore, we would be better off planning for the worst—war—than being caught off-guard.

A significant probability and some controversial photographic evidence suggested that the aliens already have a moon base. Photos taken by astronauts, some 122 of them, suggested an alien presence. This spurred on plans for our floor and the Space Defense Initiative in 1981.

Military surveillance satellites became unannounced cargo for some missions for NASA.

THE U2 PROGRAM AND PROJECT CORONA

LAUNCHING NIGHT VISION

Corso relates his initial meeting with Dr. Fredericks, a researcher at Fort Belvoir, to present the plans for developing night vision devices. Corso's name was to be erased from the appointment book, and their meeting "never took place."

Dr. Fredericks would commit to developing the night vision devices and “slip” the project into an existing contract to draw little attention.

The goggles happened to be the first project to come out of Corso’s initial efforts in his new position in 1961. By 1963, Martin Marietta Electronics, a subsidiary of Lockheed Martin, would begin distributing night vision goggles in Vietnam and Europe.

THE U2 PROGRAM

The U2 surveillance plane, or "Dragon Lady," from Lockheed Martin, monitored Soviet missile development and searched for landings or crashes of extraterrestrial aircraft.

The problem: the U2 was difficult to fly and could be shot down. Pilot Francis Gary Powers flew over Soviet missile installations at high altitudes, deliberately making himself a target to test the Soviet's abilities. His plane was not shot down, but his aircraft crashed for some unknown reason, but he managed to eject soon enough to be paraded before the media as a spy.

The solution: switch to a surveillance satellite, christened "Corona," to be snuck aboard NASA orbiters. The photos from Corona offered a better resolution, and we could scan more territory to include South America, Africa, and Asia.

Corso convinced himself that the aliens simply did not have the "manpower" to win a global war. The aliens studied us, hampered our electronics and communications, infiltrated us, but did not have the capability or intention—at least not yet—to claim the planet as theirs.

We needed an effective moon base from which to launch counterattacks without creating world panic.

PROJECT MOONBASE

"Project Horizon" would create a lunar military base to counter the aliens who already had a base there. NASA had taken over the space program in 1957, committed to demilitarizing space, which must have been great news for the aliens.

Corso states that the aliens had been blatantly violating Earth's airspace for decades, if not centuries.

We were well behind the Soviets in the space race at that time, and they put projects into orbit with amazing ease. Fearing the prospect of an authoritarian Russia in space making some kind of treaty with the aliens at our expense, we had to catch up. We needed to fulfill Project Horizon before the Soviets built a base. (Author's Note: A recent NASA-Space Force agreement promotes a base on the moon. [83] In June of 2021, China and Russia announced plans for a joint moon base. [84])

Summary of the Horizon Project documents:

- Construct a fully armed, permanent, expandable, and independent outpost with long-term survival, surveillance, and scientific research capabilities for up to twenty people—before the Russians-- who had plans to declare the entire moon as their territory.
- Be prepared for hostile aliens while building the base.
- House sufficient military power to deter war with aliens or Earthlings.
- Take a Manhattan Project approach in building the base run by a unified space command. Project Horizon would prove to be bigger than the Manhattan Project.
- Develop an earth-orbiting station (ISS?) to serve as a rendezvous point for refueling, increasing, and exchanging payloads and relaunches.

NASA, with control of space exploration, stopped Project Horizon before full funding of development. Most of the civilian sector, and their funding sources, seem set against it. Such a project would suck up scarce funding that other people wanted.

Corso managed to get Bobby Kennedy into a meeting to persuade his brother to pursue Project Horizon. President Kennedy soon after that announced a staffed landing on the moon by the end of the 1960s. Corso asserts that Project Horizon never came to fruition during his lifetime.

THE INTEGRATED CIRCUIT CHIP

At the behest of General Trudeau, Corso turned his attention to the charred semiconductor wafers separated from a bigger device. After a decade of failure, he told Corso to show them to Alamogordo's rocket scientists and Bell Labs, which had already developed high-tech transistors from the Roswell artifacts.

Corso proceeded but first got Professor Hermann Oberth's opinion on the matter. Oberth said that some members of the military-industrial complex had already seen the wafers and some of the other artifacts immediately after the crash and before they were shipped off to Fort Riley.

The Roswell craft revealed no wiring, and Oberth and Corso wondered if wafers embedded in the ship served as a solid-state electronic nervous system transmitting commands and signals.

Corso next gives a historical account of vacuum tubes and the first clumsy and bulky computers and calculators, which helped defeat Hitler by breaking his secret codes.

Bell Labs new junction transistor, a miniature silicon chip, needed little power or size to operate efficiently. It could run from batteries, paving the way for personal computers and cell phones—and new weapons systems, including guidance systems for missiles.

Successfully challenging the EBEs in space was not feasible until the Roswell crash's evolution of integrated circuits and microprocessors arose.

THE LASER

Corso drove, while trying to lose any tails, to Fort Belvoir. Corso recalled the weapon at the crash site that utilized alien technology to emit stimulated-energy radiation at high power. They also found a flash-light-sized device that emitted a concentrated light beam to render precise cuts and shine at great distances. It could even cut metal.

This technology went directly from Wright Field to contractors. The microwave oven would soon follow. Masers would emerge from Bell Labs, then the revolutionary lasers.

The laser could serve as a weapon against the EBEs but had numerous applications that transformed entire industries, including communication, weapon targeting systems, navigation, power transmission, data storage, and illumination.

Intense and thin laser beams allowed for precise surgery with minimal trauma to tissue as experienced by abductees, as evidenced in "medical professional" cow mutilations.

Corso's list of applications for lasers for the Army:

- Surgical or cutting tools.
- Rapid-firing weapons were capable of long-distance attacks.
- Targeting systems
- Signaling
- Range-finding

- The projection for large and bright screens
- Mapmaking
- Navigation

THE ANTIMISSILE MISSILE PROJECT

Corso recognizes the "Antimissile Missile" as one of the vertebrae of the backbone of his illustrious career that he approached with humility but also a great deal of pride.

In 1962, General Trudeau and Corso huddled up and catalyzed from the Roswell artifacts a laser-based system to bring down missiles in mid-air, the "Antimissile Missile. In 1963, Congress approved $300 million to develop an antimissile program. Defense Secretary Robert McNamara stood against it, enraging Senator Strom Thurmond, whom Corso now worked for as an aide. McNamara was no pacifist; he had ordered the firebombing of Tokyo. Corso suspected that McNamara had fallen under the CIA's influence, wittingly or unwittingly, and hence the KGB.

Russia could not defeat us militarily, so they toppled us from within by influencing our policies to be in their favor.

Corso became intimately involved in the hearings regarding the missile defense system and found himself authorized to disclose some of what he knew about Roswell.

From the autopsies, he knew that the EBE brains emitted long, low-frequency electromagnetic waves. This suggested that the lobes of their brains benefited from integration with electromagnetic currents. They flew their ship partly with their brains and plug-ins for their hands.

Trudeau and Corso believed flying missiles could be targeted through such a system as well. Human brains integrated into a proper system could target and bring down a flying rocket. They also thought the Russians pursued the same capabilities.

Corso included in his report to Trudeau the following characteristics of a potent antimissile system based on laser weapons:

01 Capability to lock onto an incoming missile despite evasive maneuvers. Ability to utilize satellite triangulation information.

02. House circuits resistant to heat, radiation, explosions, and electromagnetic pulses from atomic particles, for example, from nuclear bombs.

03 Show potential compatibility with proposed targeting systems arising directly from the Human brain.

Current ICBMs could not measure up in a variety of ways. We needed new types of weapons and support systems.

Corso suffered a personal hit to his psyche when funding was denied for an antimissile sky net but took hope when it was promised in a few years after we had launched enough satellites to surveil Russia and test their capabilities completely.

"A few years" turned out to be twenty, and it took President Ronald Reagan, who mentioned the possibility of hostile aliens three times in speeches, to get the appropriate funding it deserved and get it done.

FIBER OPTICS

The Roswell craft showed no conventional electrical wiring. Instead, single glass filaments permeated the ships' panels. The systems were sensitive to the wavelength of light transmitted, carrying several communication information all at once.

Fiber optics existed only on our drawing boards, but what Roswell seemed to have promised to be a sophisticated version of it. Mastering it had evaded researchers from 1947 to 1962 and would take decades more to put it to its full use, including telephone communications.

Mastering the physics of fiber optics seemed critical to learning space exploration.

The value to the military would promise to be immeasurable, giving them a huge advantage in combat.

Bell Labs remained the leader in fiber optic research, and Trudeau and Corso would pitch to them for manufacturing.

SUPERTENACITY FIBERS

Corso possessed a fibrous cloth-like material resistant to being cut by razors. Stretchable, twistable, it would bounce back to its original forms. Corso likened it to the filaments in a spider's web, which employ super-compressed molecules in an alignment that runs lengthwise. The "cloth" comprised the outer skin of the EBEs' spacesuit.

Monsanto stood out as the industry leader in developing such materials, but, in 1962, they claimed they were years away from success.

Corso's team envisioned multiple uses:

- Outer covering for craft and vehicles
- Bullet-resistant armor and clothing
- Parachutes
- Protective "skin" for about anything

The famous physicist, an eventual Nobel Laureate for his work on quantum mechanics, Richard Feynman, already had begun theoretical work on such materials. Feynman knew some of the Alamagordo scientists exposed to the Roswell artifacts.

Corso presented to Monsanto and DuPont, and DuPont patented Kevlar in 1965 and brought it into the mass market by 1973.

DEPLETED-URANIUM INVISIBLE ARTILLERY SHELLS

"Stealth" meant invisibility to radar. We had stealth aircraft, but artillery shells lent themselves to track.

Corso's team was working on the invention of artillery shells covered with such material to render them invisible to radar.

The material, likely to be ceramic, would not only have to be immune to explosively high barrel pressures but would be tasked to maintain its travel trajectory.

They needed such heavy and dense metal that it could pierce most all conventional armor with ease.

Depleted uranium fits the bill perfectly, and they hoped it could penetrate EBE craft.

They never quite perfected the outer ceramic covering. Still, the depleted uranium shells, guided by range-finders based on laser technology, did quick work out of Iraqi tanks in the Gulf War.

HARP-THE HIGH ALTITUDE RESEARCH PROJECT

The HARP artillery monsters could launch their shells (or even small satellites) into space as well as into enemies at long distances.

Dr. Gerald Bull developed the HARP in Canada, but the United States favored rocket-launched space vehicles and guided missiles.

Bull turned to other markets, and especially to Iraq and Iran. He got a buyer: Iran. Bull's decision may have sealed his violent fate as he was soon gunned down in Belgium before the Gulf War began.

MORE PROJECTS

- LIST OF OMISSIONS: Corso's team figured that what they did not find at Roswell offered implications for our sojourns into space in the future. On the Roswell craft, no nutrients or nutrient sources, no refrigeration, no water, and no waste disposal mechanisms could be discovered anywhere. No medications and no extensive medical treatment abilities other than "bandages" were present. The EBEs did not seem to need sleep because there were no beds or resting places. Assuming that the little cosmonauts did not need to return to a mothership for anything, then the EBEs brought a new meaning to the term "Spartan existence."

- IRRADIATED FOODS

The military already applied lethal radiation doses to preserve food without refrigeration, freezing, smoking, or canning. Corso witnessed and sampled one such food pantry at the Quartermaster Center.

- PORTABLE ATOMICS

The military already had portable nuclear power for energy near Sundance, Wyoming, and in Greenland at Camp Century, only nine hundred miles from the North Pole as part of the Army's Nuclear Power Program.

The military needed bases in remote areas to surveil for EBEs. Those areas would be their first choice to establish bases, if not under the ocean, and nuclear power, albeit with risks, could supply them with plenty of energy. These bases and the nuclear power that ran them were in response to what was recovered at Roswell.

TESLA'S DEATH RAY" AND THE ACCELERATED PARTICLE-BEAM WEAPON

Corso informs us that the alien craft mostly likely weaponized themselves with the ability to fire directed energy consisting of powerfully excited electrons. Nikola Tesla had been the resident expert on such "particle beam" weapons, all in theory, in the 1930s. When he passed away, one of the greatest minds and inventors of all time, the government confiscated his notes.

Since it was mostly lightning that brought down the craft, a form of directed energy, the Army figured that they might, indeed, be vulnerable to directed-energy weapons themselves.

Our first particle beam weapon underwent testing in 1958 by the Advanced Research Projects Agency or ARPA, which was only one year old.

The particle beam, whether composed of electrons or hydrogen atoms, would move close to the speed of light and energize the atoms in the target to explode. If some part of it did not explode, it would be at least unable to function electronically.

ARPA morphed into DARPA in the 1970s, and the ARPANET became the internet in 1974, and the ability to add graphics and hypertext gave birth to the World Wide Web in 1990.

After explaining the full history of ARPA, Corso next treats the reader to a full of history of Tesla. One can't discuss Tesla without addressing his archrival, Thomas Edison.

One real takeaway: Tesla lived and was created as a one-of-a-kind genius.

Since the alien craft flew surrounded and protected by electromagnetic and gravity waves, a particle beam weapon was the only way to bring it down.

The United States finally got its particle beam "railguns" functional in the 1990s as part of the Star Wars defense net.

A prototype particle beam weapon, using hydrogen atoms, found itself launched into orbit in July of 1989, some forty-two years after Roswell from White Sands Missile Range. [85]

Corso and Trudeau would both be retired by the end of 1962, but first, they had to sweat the Cuban Missile Crisis as well as hostile aliens abducting people and trying to dismantle our space program. Corso played a role in getting surveillance photos of the Russian missiles en route to Cuba by sea to the right people to try and get Kennedy to take action, who, amazingly, had left Washington, DC at the time.

Kennedy finally set up a naval blockade around Cuba. Crossing it meant war. The Russian Navy backed down and turned around.

General Trudeau used the opportunity of the Cuban Missile Crisis, as he retired, to make an even stronger case for some sort of space defense force and claimed he had viable plans for weapons to shoot down missiles in mid-air.

But it wouldn't be until the 1980s that Trudeau and Corso would have their most cherished wishes not only come true but, as we have already discussed, their fruition would lead to the end of the Cold War. The interactions between Reagan and the CIA and Gorbachev and the KGB regarding the common alien threat led to the fantastic result. The Cold War had been forced to play a secondary role to the risk of a Star War. Reagan promised to protect Russia from the aliens, and Russia pledged not to fire missiles at the United States. The Strategic Defense Initiative squashed the massive buildup of nuclear arms. In a way, the alien presence, an existent threat to Earth, may have prevented a nuclear war between the United States and Russia. General Trudeau told Corso to go ahead and publish his book in 1989, releasing Corso from their oath of silence, and finally passed away in 1991 at the age of eighty-nine. Trudeau did manage to see the activation of the Star Wars defense (SDI) and the end of the Cold War. In July, Corso passed away in 1998, just after the forty-first anniversary of the crash at

Roswell. Both men were humble American and global heroes, part of a handful of intelligence operatives who made the right calls after that dark and spooky lightning-filled night in Roswell in 1947, which would change life on Earth forever.

The government did make a subsequent statement that said that the 120 sightings of the AATI program were not of United States military design or military projects.

Do the Navy's shared sightings prove that aliens visit us? No. It could have been China or Russia, no matter how unlikely. They could have been unmanned and robotic.

Assuming the film is real, plausible explanations are exceedingly rare for footage from the US Navy showing such astounding maneuvers. If it looks alien and moves like it's alien, and nobody is claiming responsibility, it's alien.

We will never forget the famous "weather balloon" of Roswell. My father heard the announcement of a recovered flying saucer on the radio, which was retracted the next day.

Can we trust the government to tell us what we want to know?

The most convincing evidence the government released has to do with the bizarre materials in their possession. They allowed a contractor civilian, Eric Davis, to make statements about these materials being not created on Earth, a way to inform the public that UFOs are real, that aliens have visited Earth, and save face at the same time.

Through the Freedom of Information Act, Luis Elizondo received a letter from the US government stating that the Pentagon tracked UFOs at least since 1950. [86] They are most interested in sightings around nuclear missile facilities. Corso said they tracked numerous UFOs for a few nights preceding the famous crash back in 1947.

EVIDENCE ALIENS SEEDED LIFE ON EARTH THROUGH DNA

Vladimir shCherbak and Maxim Makukov published an article in 2017 in the International Journal of Astrobiology concerning "Biological SETI."

Given that some scientific literature proposes that life on Earth might have descended from seeding by an ancient extraterrestrial civilization that became motivated to spread energy, a testable hypothesis arises. Microbial seeds sent into space could carry a signature that could endure for extended periods. Furthermore, this signature would be encoded in genetic material since it allows for long durability periods with integrity and is the least evolving component. Therefore, such a signature might be found in the genetic codes of the organisms of life.

In 2008, part of Vladimir shCherbak's work on the mathematical nature of DNA appeared in The Codes of Life on pages 153-185 as The Arithmetical Origin of the Genetic Code.[87]

DNA, according to shCherbak, encodes a privileged numerical system. The genetic code, he believes, is a syntactic arithmetic structure that probability dictates are artificial—it did not arise naturally.

The genetic code used for its arithmetic is the decimal place-value numerical system that included a conception for zero. The zero sign governs the genetic code as a necessary part of the decimal system and an acting arithmetical symbol.

Therefore, the genetic code displays an artificiality because of the non-material abstractions, the concept of zero, a syntax based on decimals, and unique summations.

These findings suggest that the genetic code did not arise from chemical evolution on Earth but instead represented pure and elegant mathematical information from elsewhere.

In May of 2013, scientists Vladimir shCherbak and Maxim Makukov published a paper entitled The "Wow! Signal" of the Terrestrial Genetic Code in which they suggested that DNA is proven to be embedded with a variety of sophisticated coding and activation keys which includes mathematical constructs, not the least of which are decimalization and an expression of zero. DNA, they say, first arose from artificial engineering, agreeing with Francis Crick, and bears cleverly encoded mathematical and pictorial signatures of its makers.

SECTION: ANCIENTS

ALIEN DOMINANCE OF EARLY CIVILIZATIONS AND RELIGION

Current Christian theologists turn to Genesis to tell the faithful that a single all-powerful God, although judgmental, temperamental, and often ruthless, created the universe. They do not tell us exactly how. They lecture us about the origins of Humans and the original sins of Adam and Eve.

Paul Wallis, an experienced clergyman, lecturer, and author, took a close look at the book of Genesis and wrote Escaping From Eden. He had one thing going for him. He had plenty of free time. He laid flat on his back from an injury and could not preach for a while.

He came across such anomalies and contradictions that he concluded that he could not be reading the original versions of the famous stories.

His first task is to find the original narratives and clarify their suggestions about where we came from, who we came from, and who we are.

Ancient stories of human origins were nearly eliminated from the Hebrew Scriptures in the sixth century BC. Instead, Yahweh would eventually appear out of nowhere and select the Israelites as the chosen people. How convenient.

Many creation myths found themselves denied access in the 2nd and 3rd centuries AD in the Christian writings.

Why?

What were the new authors and editors hiding as they crafted their self-serving myths to establish one God and church as permanent rulers?

As we take the same red pills that Neo took in the *Matrix* and join Alice in going down the rabbit hole, we find ourselves roaming through ancient Greece, Africa, India, Mesoamerica, and Sumeria. According to the myths, we are not searching for a traditional monotheistic figurehead but, according to the legends, for evidence of a Sky Council—a heavenly council of extraterrestrials. Regarding Genesis, we will have to discard the illusions of a monotheistic god from the past to embrace new possibilities of Human projects orchestrated by advanced aliens. For a Darwinian milking of a planet, various alien species would compete to access resources, including the human genome and humans themselves as biological booty.

History is clear in the thousands of cuneiform tablets in Sumeria. Early Human civilizations constructed impressive buildings, puzzling myths, mined resources, and fought wars. They did not, however, render artistic images depicting only one god or one humanoid form.

Playing explorer, researcher, clergyman, and detective, Paul Wallis in *Escaping from Eden* undercovers mindboggling secrets about Biblical texts. Their implications demand a rethinking of the Abrahamic religions and a new openness to alien contact, influence, and even responsibility for creating Humankind. His conclusions will be as controversial as informative and broad-scoped as their range of challenges to entire belief systems.

Not only is Genesis a false narrative, but it also turns out that our genetic code and its DNA orchestrator of most life likely originated off-planet by advanced alien genetic engineers and came here through space over four billion years ago. That civilization may be our "makers" and might have paid us a few or many visits to tweak us in the direction *they* wanted. It looks like they encountered some competition along the way.

The entire universe and the origins of Humans come into an improved—but unsettling-- scientific and spiritual focus.

Genesis 1.1

The Hebrew and Greek of the Septuagint versions of ... ible refer to God as "Elohim," a plural noun, implying more than a single God.

Genesis 1:26

"Let us make?" as in "Let us make humans look like us-in the images and likenesses of ourselves?"

The New Jerusalem Bible suggests that "image" implies a physical resemblance, like the son of Adam and Ada. "Image" reflects a term that is "concrete."

Again, we have a plurality, "us," and not a single God.

Genesis 2:10-14

Why does the described garden of Eden happen to reflect high-grade deposits like gold, onyx, and resin? Who would make effective use of those materials?

Genesis 2:17

The Tree of Knowledge is forbidden.

If God wanted humans to love and have free choice, why deny them the knowledge of evil versus good, which seems a prerequisite for choosing mates and a destiny? Is not knowledge useful for survival? Did God rig life for failure? Picking from the tree of knowledge is punishable by death. Does this sound like a loving and just God?

ـ 2:21

_od names the local fauna, but which offered up no suitable helpers. Seeking a capable helper, God put the first "man" into a deep sleep, extracted a rib, and used the flesh to patch him back up. From the rib, he fashioned a helper, "woman."

Is the all-knowing God short-sighted?. God, the ultimate source of all wisdom, forgot to make women. Much to modern woman's chagrin, women earned the subservient role of "helpers" only after animals were deemed incompetent.

Genesis 3:1 Now the Snake...

But the first man has arms and legs, has a brain that can handle language. "Man" does not include talking snakes. But an intelligent and talkative snake shows up out of nowhere. The Snake has a one-track mind to convince Eve to partake from the Tree of Knowledge, and, of course, she does to become conscious.

(Author's note: Genesis 1:24-27 teaches that man came into being after the animals. Genesis 2:18-19 days before.)

Genesis 3:16 Women will give birth, but with lots of pain.

Is this a mere consequence or punishment? There is to be no childbirth without it. All of that pain will come after consciousness, sexuality, and self-awareness.

Genesis 3:22 And Yahweh Elohim pointed out that the human became similar to one of the—plural. The word "God," again, represents a group.

Genesis 4:14 Cain is driven from the Garden of Eden: He will be a wanderer, restless, on the Earth, and whoever finds him will kill him. If Cain wanders beyond the garden, he is a goner. But who will kill him? It can't be Adam and Eve's children.

Genesis 6-9: The animals pair up.

Wallace points out that it wasn't that easy. Another part of Genesis implies a much more complicated boarding system.

Genesis is a hodgepodge from a variety of sources. Some refer to "Elohim," others to "Yahweh," and Moses becomes another version of a narrator "J."

We cannot avoid a major, mindboggling conclusion. The use of the word Yahweh in Genesis reveals that we are not reading and studying the original version. The author is altering the story as he or she sees fit and leaves a trail a blind person could follow.

What if the original version wasn't about Yahweh at all and instead spoke only of the Elohim as a plurality. Then, Genesis, he claims, might not reflect an all-powerful single God at all!

Genesis 6:1-4 The Nephilim...the giants.

All living creatures were said to have been destroyed by the great flood—all except the Nephilim. Were the Nephilim good swimmers or boat builders?

The Bene Elohim, not purported to be giants themselves, nevertheless had giant offspring when they mated with human daughters, who all happened to be beautiful. If human daughters were somewhat of a sexually desirable novelty, who were the rest of the women at that time, and where did they come from?

The Bene Elohim are not believed to be human. Some claim they were angels, but Hebrews in the New Testament deny it. According to Wallis's reasoning, the bene Elohim was neither God nor human. Wallis suggests they were aliens.

Genesis 20:13: Blame it on the Elohim. These mischievous Elohim caused him to wander from his father's house. Abraham applies a plural verb to describe the actions of a plural Elohim.

Genesis 35:7: Jacob built Bethel and its altar because of the appearance of the Elohim (plural usage again) when he fled from his brother.

Genesis 11:7 Yahweh refers to himself in the plural sense.

Genesis 22: Abraham sacrifices his son, Isaac. We stand puzzled by what a loving God tells Abraham to do. On the one hand, God promised him a family through his son but then told him to commit the ultimate pagan sacrifice and kill his son. An Elohim different from Yahweh makes more sense, and that they took different sides in bossing Abraham around.

Author J shares little about Elohim but writes as if their Nature is common knowledge at that time. Wallis asks about the other sources that J believes should be on our library shelves.

Because there are simply too many gods in these sections, J makes a feeble attempt to tidy up all of the lost strings and return towards monotheism; J translated Elohim to be the same as Yahweh. The reader is thrown into a serious problem from which there is no escape. God demands obedience that bears no questioning, demonstrated by asking Abraham to murder Isaac, turning God into a polytheistic monster.

The Elohim threw a monkey wrench into the story of Genesis to such a degree that if it cannot logically be explained away, the teachings of Genesis, he offers, should be avoided altogether.

What is particularly troublesome is the apparent labors and influence of multiple gods rather than one. Humankind, they said, looks like one of "us."

Interpreters propose that God consulted some heavenly Council for advice and suggest they were angels.

One problem, the authors of Genesis, forgot to allow for the creation of angels, period.

In Psalm 82:1, the Elohim are suggested to comprise the heavenly Council. In Psalm 89:6, the Elohim reside in the skies with God can be compared to him, although inadequately. The Elohim stand accused of harboring a member referred to as "the accuser." In Job, the word accuser is translated as "Satan."

If in Genesis 1, Elohim is plural, then Elohim didn't have a heavenly council; Elohim was the heavenly Council. A strict translation of Elohim is "powerful ones." Plural, plural, plural.

In an attempt to bring harmony to the stories of Yahweh and the Elohim, J and friends instead fostered moral chaos and dissonance.

Were these authors trying to hide the fact that the themes of Genesis arose from the actions of lecturing from a group of leaders, a form of Elohim council of aliens, rather than a monotheistic god?

This entire analysis suggests that Genesis is an altered version of someone else's sacred story to install just one ruling God.

NON-HUMANS VS HUMANS

Once Wallis discovered that Elohim was not only a plural term but also represented some alternate possibility for God as a ruling council in Genesis, he felt like Neo in The Matrix, realizing that his old world reeked of illusion and that his new world offered infinite bouquets of new realities .

At the time, Wallis, a clergyman, found himself physically incapacitated with an injury, so he would have plenty of extra time to reevaluate his understanding of God and his role in Humanity.

He also realized that Adam could also mean soil or Earth, or earthling. Stories about Adam became myths about earthlings or Earth. The word Eve told the living before being assigned as the name of the first woman. Stories about Eve became myths about the residence. These two words hold themselves as integral to orthodox Christianity and Judaism.

When Wallis analyzed the word Elohim similarly, he found himself forced to venture back to the seeds of Genesis to discover an ancient story from a distant land predated the Bible.

Genesis

In the beginning, the barren and lifeless Earth is covered in water and darkness. Powerful Ones, or spirit-beings, orbit above the water. Light is born and manifests the sun, stars, and moon.

Wallis conjectures about superhuman powers that would have to be possessed by the Powerful Ones, a term he will use to mean the Elohim.

(Note: Science shows evidence of the Big Bang and how the stars, our sun, and moon were formed after it. See my book on The God Equation by Michio Kaku.)

Genesis credits the powerful ones with creating a habitable planet teeming with life and natural resources. Some of the elements of the Earth form a lump of clay and become the building blocks for humans to be made in the Powerful One's image. They make women so the men won't be lonely and depressed. The Snake pops up out of nowhere and shows the need for explanations that would remain for thousands of years until the author(s) of the New Testament tried to clarify.

(Note: The first DNA is shown to exist about four billion years ago on Earth. Genesis claims that the Powerful Ones orchestrated all life on Earth, which took four billion years of natural selection to achieve. The seeded hypothesis states that genetically engineered DNA came to Earth through Panspermia from an ancient alien civilization four billion years ago. As a central theme, we will find similarities in abilities and results for the Powerful Ones and this ancient alien civilization, which may have visited Earth.)

Job 1.6 likens The Snake to a bene-Elohim spirit being and sits on the heavenly Council with his fellow Powerful Ones. The Snake, in the vein of Satan, creates chaos and conflict. Women urge Humanity to accept temptation and become self-aware, and put on some clothes. Brain-dead is replaced by understanding, sexuality, and morality. In which pain is mandatory, childbirth is part of the price women must pay for their sin of acceptance.

This progression catalyzes a war among the Powerful Ones of the Heavenly Council. The war reflects different agendas and wills, so much as to persuade Wallis to drop the word "heavenly" from "heavenly council." The Council is anything but lovely, divine, peaceful, and beautiful.

Incoherence arises from translating the Council into a monotheistic God who falls short of an all-knowing creator and expert in the universe. He overlooked the need for men needing women to keep them company. While good himself, with a strict morality, he creates men with no morality at all. How could he blame a man for an immoral decision if he has no morality? Why should achieving a moral level of consciousness be punishable with the pain of childbirth and life replete with hard labor that ends in death? How could an all-knowing God create The Snake and not anticipate the problems that would stem from that creation? God even has second thoughts about creating humans at all!

Wallis concludes that lumping the Council into a single "God" results in such moral distortion that God appears fickle, sloppy, wrong-footed, and ruthless. Suppose we, instead, view God as represented by a council of competing individuals, the Elohim or Powerful Ones. In that case, we can blame God's shortcoming as the result of infighting and other wills and agenda. With Satan himself on the Council, who could not see this coming?

Satan has led one-third of the Council, and they find themselves banned from heaven to be exiled on Earth. Satan upgraded the humans without a majority vote. He "broke rank" and suffered the consequences of his actions—or did he? He and his friends immediately have sex with human women to product giants deemed titans and demigods. This myth permeates countless cultures, including Greek, Roman, Hindu, Norse, Celtic, and Mesopotamian. The demigods go by different names, but the story and meaning remain the same.

(Note: Since we are talking myth here, the devil, sex between deities and humans, and giants kind of myth, it looks like someone, early on, told a compelling story. Everybody else plagiarized it for thousands of years.)

As human populations expand, so do management headaches. Humans live too long, but Council members stand accused of abducting human females to impregnate them to bear hybrids. This was not part of the main plan. The solution: kill everyone with a devastating flood.

(Note: In ufology, the aliens stand accused of abducting people in the millions to make hybrids. Again, we see another similarity between the Council and advanced aliens.)

THE MYTH OF NOAH

Noah is selected, although we have no clue about the process, to build a boat, fill it with enough plants and animals to continue life on Earth, and plan to repopulate the Earth with the DNA of his extended family.

Similar to the Satan and giant demigods myth, the Noah and the Great Flood stories permeate many world cultures: the Aboriginals in Australia, the first Native Americans, and the Philippines, Polynesia, Finland, Turkey, Hawaii, Iraq, Malaysia, Ireland, Korea, Peru, and China. Again, a catchy and intriguing, and lesson-giving myth becomes plagiarized about the world. If you are going to tell remarkable stories, Noah and the Great Flood simply cannot be passed over.

The Council changes the meal plan, adding meat and fish. The moral court is held to judge murderers, and animals find themselves in court as well.

The new society travels east and settles on the Shinar Plain, Mesopotamia, to begin the first civilizations in Babylonia and Sumeria. These new civilizations were said to be gateways between earthly humans and the heavenly Council.

Wallis believes that El is an abbreviation for Elohim, and that leads to the conclusion that the word Babel could be construed to imply for The Powerful Ones or a gateway to. The city of Babel had a portal or a wormhole to the heavens.

The Council pays a visit to what they had seeded and decides that the humans morphed into a civilization that bears too much capability. We might guess that they felt threatened by the humans' progress. The Council decided to halt the achievements by completely leveling the new and prosperous city-states. They even snatch away the common language the humans were proud to develop.

Genesis 12:1

Abraham gives us a tour through a world reminiscent of Earth's time of nomadic tribes, settlements similar to city-states, and the start of empires. This world reveals peace treaties between wars and world trade. Negotiation amidst diversity best describes its culture, and religions flourish.

Wallis embedded his inquisitive mind in myths older than the Bible, including the stone tablets carved with the cuneiform script of Sumeria of ancient Mesopotamia. The stories validate Genesis—if we take Elohim to imply plurality or a collective.

HAVE I SEEN YOU BEFORE?

Wallis describes the discovery and copying of some cuneiform script over 5,500 years old. Ancient Mesopotamia served as an archeologist's mecca of over 200,000 clay tablets interred in the ground. The tablets found themselves discovered by archeologists in the 1500s but were set aside until the 1800s. Henry Rawlinson determined that the stones contained some form of a proclamation from the royalty that would be called the Behistun Inscription in Elamite, Akkadian, and Persian languages.

The tablets spoke of the lives of Babylonians, Assyrians, Akkadians, and Sumerians. They also repeated myths of even earlier times, Sumerians, many of which persisted into other cultures. One story concerning creation speaks of powerful beings in collaboration rather than a single "god." These beings became known as the Sky People. One of the myths parallels the Christian myth of Adam and Eve in the Garden of Eden, but with a few twists and turns as far as who creates who. There are also similarities to the one with nature theme we see in many cultures, from the Australian Aboriginal walkabout to Edgar Rice-Burroughs Tarzan, to Mowgli from Rudyard Kipling. They all have the same ending; the protagonist ends up civilized and often has the help of a beautiful woman.

In the cuneiform texts of Sumeria, the Sky People (aka Anunnaki) exploit their new creations as farmers and servers of sacrificial food. We encounter the same Sky People/Elohim fear of the human population growth, advancement, and a great propensity to foster management headaches. Mating the Sky People with humans is particularly disturbing, leading to such disharmony that scheduled the proverbial world flood.

These stories, in cuneiform, predate their biblical plagiarisms by some three thousand years! For example, Instead of Noah, we have Ziusudra in Sumeria, Atrahasis in Arkadia, and Utnapishtim in Babylonia—all saving Humanity with an ark and other embarrassing similarities. If there had been ancient copyright laws, they would have been broken. They all end with a trip east to build cities like Babylon and Babel, the stargates to heaven.

The authors of Genesis plagiarized the Sumerians and Psalms, Ezekiel, Luke, Isaiah, Job, and even the Book of Revelation in the New Testament— a war in the heavens.

Wallis concludes that the cuneiform myths served as the literary source for Genesis or that Genesis and the cuneiform myths both arose from the same even more ancient origin. Based on the assumption that Sarah and Abraham were already seniors upon leaving Sumeria, Genesis patterned itself after the Sumerian version, translated into Babylonian and Assyrian versions. The authorship of the Bible comes under suspicion, totally discredited by this claim of a copycat Genesis. Became heresy. Now we can understand why the tablets might have been discovered in the 1500s but remained hidden to the world for the next three hundred or so years.

Wallis witnessed few temper tantrums among religious scholars when he conveyed the similarities between the Sumerian texts and Genesis—until he pressed the issue that the Sky People, alluding to extraterrestrials, engineered Humanity rather than the stereotypical monotheistic God.

The Christian Bible, the ultimate authority in morality, behavior, and Faith, arose from a false version of older myths. The validity of the Christian Bible became dependent on the invalidity of the Sumerian myths, which had predated the Bible for thousands of years. It seemed obvious to Wallis that falsehoods and telltale inconsistencies jumped out from the pages of Genesis to an educated reader. Faith, guiding huge populations, Wallis concluded, should not arise from turning lies into authoritarian truth.

Wallis reasoned that if the Sky People frequenting the cuneiform tablets were the same as the Elohim or Powerful Ones of the Bible, then Yahweh must not be viewed as a collective of Powerful Ones but as a unique and separate entity that intermingled with the Powerful Ones.

Joshua 24:14: The Israelites are lectured for doing the bidding of Yahweh, who had been revealed to Moses. Moses is succeeded by Joshua, who informs the Israelites to dispatch the Elohim, in plural form, and serve Yahweh instead.

Unless God has a multiple personality disorder, Wallis states that Joshua tells us that Yahweh and the Elohim cannot be equated—a smoking gun of Biblical proportions.

The Israelites fled the Sky People to commit to Yahweh for their safety just as the ancient Egyptians fled an old regime under the influence of Powerful Ones.

Wallis continues the argument that the Sky People and Yahweh are different by presenting a statement that the Sky People reigned as individual kings for as long as 36,000 years if we can believe the Sumerian concept of time.

The superhuman, long-living Sky People, according to the myths carved into the cuneiform tablets, proved to be capable of:

01 Living in the heavens, in spaceships or stations.

02 Controlling the weather.

03 Terraforming barren land into fertile land.

04 Genetically engineering human beings

05 Applying sophisticated technology.

06 Political maneuvering.

07 Breaking rank.

08 Debate.

09 Participating in power struggles.

10 Scientific experimentation.

11 Abducting and exploiting humans.

Wallis is convinced that if we assume that the tablet myths represent some truth, the Sky People could be little else other than extraterrestrial beings. For confirmation, he sought out paleontologists, anthropologists, and archeologists, hopefully, who would not be influenced as much by religious beliefs.

MAXIMIZING AMNESIA

Wallis cites Genesis 6, in which there are references to two extraterrestrial species, the ben Elohim and the Elohim, the human species, and the Nephilim, which are a hybrid between the Elohim or ben Elohim and humans.

A most important takeaway: All are referred to in the plural form!

The species tally matches that of the ancient cuneiform texts.

So why didn't that story become the gospel as represented by such biblical scholars as Marcion as early as the second century? Wallis attributes the acceptance of today's Bible to a combination of denial, coverup, and ridicule and censoring of the "true" versions of the ancient myths.

Wallis hopes that some evidence might be found in our DNA.

Peruvians from a culture some 3,000 years ago boasted skulls that measured 60% heavier, on average, than our skulls .

We have two parietal plates; they had only one. Their skulls receded much more than ours. Our brains are 25% smaller.

The culture: location, Paracas, a desert peninsula on the southern coast of Peru, was first discovered by Julio Tello, an archaeologist, in 1928. The skulls bore no signs of artificial manipulation; they formed naturally but showed no unusual or "alien" DNA.

Some similar skulls were found between the Caspian Sea and the Black Sea.

Wallace then addresses the complex human lineage that arose from interbreeding between Homo Sapiens, Denisovans, and Neanderthals.

He also gives some academic support for "giants" among Biblical tribes, including the Nephilim hybrids. He provides more evidence in the more modern recovery of skeletons of extremely tall beings and extremely small ones. Such diversity is present in numerous carvings and cave paintings.

Wallis ends the chapter by suggesting that Genesis 1, rather than a myth of creation, may be a story of recovery after the Clovis comet of 13,000 years ago.

A REMARKABLE RE-EMERGENCE

The post-Clovis world became engulfed in darkness and water and hosted a few barren islands populated with a few species. This description matches the creation conditions described in Genesis, Mesopotamian tablets, and the Mayan culture.

An ice age would come next, typified by some murderous "flash-freezing" establishing the extinct status for the saber-tooth tiger, the woolly rhinoceros, the mastodon, and the mammoth.

Wallis argues that the aftermath and climate change after Clovis might have killed most humans and forced the survivors into living underground in many cases. Archeology confirms the theory. Now Wallis brings forth the magic of the Powerful Ones to terraform what was left of Earth back into a more agreeable place to live—like the myths from the Sumerians and the authors of Genesis, and to teach humans agriculture.

The creation meaning of Genesis becomes, instead, one of re-creation and matches more closely the myth of Genesis 1, and that would support the theory that aliens landed to help rebuild after news of a global catastrophe.

The mythology of Popol Vuh, discovered by Franciso Eximenez, a Dominican Priest at the beginning of the 18 th century, supports this claim and refers to the miracle workers like engineers. He concludes his version by destroying the failed attempt at an acceptable human race, made with maize and corn, Mayan style. The remnants, as small, ape-like creatures, ran off to live in the forest and go all Darwinian to create humans again.

The beginning of Chapter Seven in Escaping From Eden tells us about another myth, on another continent, Africa, which includes Elohim type of beings, the Efik of Nigeria, and their Sky Beings. Their story shows a strong resemblance to the others that the Sky Beings get the humans started. Still, the humans soon become unruly and too successful, and the Sky Beings ban them from the heavens and condemn them to lives of suffering and conflict forever.

Wallis turns the tables and judges the actions of a God who is supposed to be moral, just, and perfect. God fails his test, and genocide is no longer forgiven but, indeed, mass murder. Callous retribution is another unforgivable offense. But if "God" is a group of power-wielding aliens, their actions become more understandable if not forgivable. They compete with the humans they enslaved for control of Earth.

The people of Babel fall into a destiny of genocide from making a building that became so tall as to infringe upon the lower border of the heavens.

Isn't genocide under Yahweh's rule a bit of an overreaction?

Wallis suggested a tighter translation, in which Yahweh is a council of advanced aliens. Rather than being too tall, the building is to house a stargate into the heavens—a real threat. The Elohim dispatch three hundred observers to the stars. Humans are not to enter space, and so to meet this threat, they destroy all of Babel by genetically engineering them back towards a state-centered about non-intelligible language, to become but animal-like once again.

History goes silent until Sarah and Abraham emerge from Sumeria to begin their journey. Wallis cannot find a purpose or motive for these repeating myths. Why would humans demean themselves? The Elohim become selfish, enslaving, and detestable rather than glorified. The True God earns no glory, either. There are no motivational speeches or inspiration for Humanity to come out of the Elohim/Yahweh myths of creation and destruction.

THINGLY EVIDENCE

Wallis finds himself arriving in India to try and solve the disturbing puzzling conundrum he fell into and couldn't escape.

He noticed the vimanas, the ancient Vedic scripture's version of spaceships, atop every temple, and on pillars, including some standing up amazingly well for being over 1,500 years old. Such archeological wonders would give rise to the fame of Erich Von Daniken's Chariots of the Gods.

Daniken credited advanced aliens for many fantastic achievements of early civilizations. He hypothesized that our DNA came from alien spores, and they wanted to encourage the development of intelligent life. The aliens would be standing by to assist if needed.

Another hypothesis: Daniken believed there occurred sexual relations between aliens and humans, creating hybrids that would become Homo sapiens.

Wallis looked to the work of Carl Sagan, including his novel Contact, and found some support from atheist Sagan.

NOBODY SAW THAT

The archeological find of Ruz Lhuillier deep in a Mayan Temple at Palenque—boasts an oversized and impressive sarcophagus made of stone for a Maya ruler, or ajaw, named K'inich Janaab' Pakal.

Erich von Daniken made the sarcophagus famous when he alluded to it as representative of ancient astronauts as a space capsule flying through the Milky Way with rocket controls, a complex chair, a breathing mask, and plumes of rocket exhaust.

Paintings and carvings of the vimanas of Indian depict spaceships, and the astronauts wore communication devices, spacesuits, and helmets. Similar artwork is found in the ancient cultures of the Chinese, Japanese, Aztec, Mayan, Guatemalan, the Hopi of North America, and the Aboriginals in Australia.

Ezekiel in the Bible describes a large cloud of smoke laced with lightning descending with wings and extended legs made of metal, a wheel-base also made of metal, and opposing thrusters. The pilot bore an appearance like a man but was not a man per se. In subsequent chapters, eleven of them, Ezekiel claims, rode in such craft. Ezekiel thinks, of course, that the astronaut directly represents Yahweh.

Ezekiel's stories propel him to a position as a prophet.

Wallis cites a few examples of ancient stone artifacts and structures that would be difficult to create without the precision and power of technologically advanced tools, including sonic energy, implying that ancient aliens may have played a role in their creation.

LOVE THY ALIEN

President Truman's National Security Act of 1947 created the CIA and placed Maury Island, Washington, and Roswell, New Mexico UFO incidents into the classified filing system of the government.

At Roswell, the government first said a flying saucer had crashed, becoming headline news in the local newspaper. Within five hours, they switched to a weather balloon story.

The obvious question: why classify a weather balloon crash?

Many former US astronauts and other NASA personnel explicitly stated that they either saw UFOs or believe that we are not alone in our universe. Edgar Alan Mitchell, Lt. Col Onizuka, Brian O'Leary, Buzz Aldrin, Al Worden, Eugene Cernan, Donald Slayton, Scott Carpenter, and Gordon Cooper.

Edgar Allen Mitchell claims he knows of evidence that aliens mastered zero point or free energy and believes Roswell marked an event of obvious contact and was not the only one.

HACKING NASA: THE CURIOUS CASE OF GARY MCKINNON

McKinnon, a citizen of Britain, hacked NASA, the Pentagon, the Navy, and the Army systems in 2002, which happened to have no firewall few high-level security measures at that time. He downloaded files to suggest not only contact with extraterrestrials but also different forms of collaboration. He uncovered what he thought to be a secret Naval force and a list of officers either connected to extraterrestrials or extraterrestrials themselves. The discovery of free energy and zero-point energy, he thought, remained hidden by the government. The United States wanted to prosecute McKinnon and lock him up for at least sixty years. Theresa May not only blocked extradition, but parliament changed extradition laws in McKinnon's favor. McKinnon lost his appeal in 2009 but won a second appeal and stayed. [88]

Parliament debates are public, and suddenly, Parliament members found themselves confronted by a screenshot photo of an alien craft and even the names of "extraterrestrial officers" and a suspect secret Naval force run by the United States.

THE VATICAN AND ETs

Fr. Jose Gabriel Funes, in May of 2008, a Roman Catholic Jesuit Priest, started to speak and write about contact with extraterrestrials. Funes served as the Director of the Vatican

Observatory and as a Senior Vatican Theologian. Christians, he argued, should prepare themselves for embracing and loving our alien sisters and brothers.

Pope Benedict XVI convened a world colloquium in 2009 to explore the theological repercussions of contact with ETs. The Vatican welcomed thirty theologians and scientists for private discussion and debate. Funes argued that man could not establish God's creative limits, and therefore, there is no serious conflict between Faith and astrobiology. Aliens, he concluded if they exist, they must also be God's creations.

Considering that the Church burned people at stake four hundred years ago for stating that other intelligent species might exist, this turn-around by the Vatican is quite dramatic—and puzzling.

Do they know something we don't, that contact might be shared with the public soon?

On May 9, 2001, we must also recall that many respectable and sincere witnesses assembled at the National Press Club in DC to share their experiences and knowledge about ETs.

Some violated nondisclosure agreements. Some found themselves and their families threatened with harm. Some major media players attended but refused to report it to the public.

A member of the Vatican Curia, Monsignor Corrado Balducci, assured the public that ETs are not the demons many believed them to be but are real and should be taken seriously. The Bible includes descriptions of intelligent non-humans: not demons, not angels. He added they should not even be called "alien" because any creature made and loved by the same God who created and loved humans should be welcomed as a brother or sister.

The Vatican's public revelations of both the aliens' presence as one of God's creations support the Sumerian and Babylonian teachings, including the plurality of the Elohim.

DNA GENETICALLY ENGINEERED BY ANCIENT ALIENS

Makulov and shCherbak's work is included in the section on Francis Crick, but let's look at it from Wallis's perspective.

Their work came to a pinnacle in 2013 with a publication about the alien origin of DNA. Vladamir shCherbak and Maxim Makulov are from the Fesenkov Astrophysical Institute and the Kazakh National University. They studied the Human Genome Project for over thirteen years and showed their mathematical evidence that our DNA arose from the intelligent design from alien genetic engineers. Our DNA bears symbolic language and arithmetic patterns. It is also over four billion years old. This implies that the genetic engineers had to be from a time significantly earlier.

They published in Icarus and, according to Wallis, suggested that the "junk" coding in our DNA, up to 75% by weight, is genetic code from extraterrestrials. (I could not find a study confirming this suggestion and did not see it in their abstract)

Some scientists from the University of Michigan think that one function the junk DNA serves is to hold the genome together. [89] Another portion may be dedicated to the intricate tasks of brain neurons.

Recall that Francis Crick and Leslie Orgel published a paper in 1981 hypothesizing panspermia but did not have the genomic and mathematical evidence to back it up.

Mikulov and shCherbak claim our genetic code contains virtual representations of prime numbers, rarely occurring in Nature. Amazingly, nine multiples of the number 37 occur throughout our code at a probability of one in ten trillion. DNA coding, if anything, did not arise from randomness, according to Mikulov, or natural selection, according to Francis Crick. If DNA did not occur naturally on Earth, then it must have an extraterrestrial origin. Mikulov and ScCherbak think DNA arrived, by chance, via comets; Crick thinks unmanned spaceships made the most sense.

The seeded hypothesis, also known as Directed Panspermia, inspired Carl Sagan's novel Contact, writings by Erich Von Daniken, movies Star Trek II: The Wrath of Khan, and Ridley Scott's Prometheus and Alien Covenant. DNA spores could travel by comet, meteors, asteroids, or spaceships—or even through space if encapsulated properly by radiation-resistant microbes.

The scientific community found itself so convinced of the possibility of tiny DNA-filled cosmonauts that they penetrated a comet with a probe named Rosetta in search of DNA. They discovered glycine and phosphorous, both integral to DNA, cell membranes, and protein.

Wallis begs the next logical question. If DNA arrived via aliens, might they follow up and tweak life on Earth towards higher intelligence some billions of years later?

FATHER GOD?

Scientists like Stephen Hawking and Michio Kaku would say no God was needed to create our universe and run it, or the universes that preceded it and will follow it as part of a multiverse. The universe has no gender nor judgment. See my Summary: The God Equation by Michio Kaku book.)

Wallis applies his assumptions about God to give God credit for human creation, even though it looks like humans arose from the work of ancient aliens.

Trying to resolve more of the contradictions, inconsistencies, and meanings of more confusing myths, Wallis explores the origins and implications of the Ten Commandments.

The Ten Commandments pit the Powerful Ones against Yahweh. Of course, Yahweh must win out, even though their battles are murky, to wind up with the single all-powerful God the Churches wanted.

When Wallis turns to myths in which people speak directly with Yahweh, we know we have fallen too far down in the rabbit hole to trust any information. Wallis covers the apostles and Moses. He wonders why Yahweh, the one and only God, would need some form of a spaceship to cruise by and pick up Elijah, the prophet.

Wallis is convinced that Christian texts simply imported Judaism's stories of Elohim, although not without trying to disguise it. Many other cultures did the same.

"FRIENDS NOT OF THIS REALM"

Wallis tackles the dubious myth of Jesus. He claims that the appeal of Jesus to the Hebrew Scripture confirms them as inspired, applies them as references that serve as prophecies and a source of spiritual interpretations.

(Note: When the scriptures accuse God of being unjust and savage, they don't mean it according to a strict interpretation. It's as if the "sacred texts" can be twisted and turned to tell whatever the interpreters want them to mean to maintain allure and power. The arguments between the endless interpretations of fudged myths finally devolve into "My God is the only true God" or, essentially, "My God is better than yours.")

Wallis believes Jesus is more the Son of Man than the Son of a punitive, fickle, genocidal, mean-spirited, and unjust God. Jesus represents Plato's migration of the soul, arising from a creative and cosmic consciousness, taking physical form, and returning to the non-physical. Jesus, says Wallis, meant to help people in their versions of this same journey.

In a video, Wallis hints that Jesus might have arisen from an abnormal pregnancy initiated by aliens. Yes, according to Wallis, Mary may have been an alien abductee and impregnated by aliens. (Hopefully artificially!) (Note: If Jesus demonstrated unworldly powers, then alien genes, and hook-ups to electromagnetic power, could be his source.)

The Romans hijacked a delusional but workable religion and way of life to turn it into a system of enslavement to control populations through fear and force rather than love.

WALLIS'S CONCLUSION—IDENTITY?

Wallis ponders his realization that our galaxy is populated by many more advanced civilizations than he thought and that some of them may have been seeded by, and with, humanoids.

What would Jesus say?

The words, phrases, and sentences that can be attributed to Jesus directly are exceedingly rare. Still, he once said, allegedly, that he lived in the company of others who were not "of this fold" and that he must bring them along, also, on his calling. Did "not of this fold" mean extraterrestrials?

Embracing aliens as our siblings are full of conflicts with historical activities. Aliens trespassed on Earth and stole resources, toyed with our lifeforms, played God through advanced genetic engineering, abducted people, and removed the uterus, eyes, tongues, blood, and genitalia other organs from some humans and way too many cows. Another way to look at it: humans are the "children" or "property" of ancient aliens, who are not aliens at all, but the original creators of life on Earth or visitors from lands far, far away, who still may share an evolutionary history with us.

Perhaps "aliens" is just not a good term at all to describe our visitors or our "creators." "Extraterrestrials" is the best term so far.

Was all of this OK with God and Jesus? Wasn't Jesus of alien genetic code and DNA, too? Flesh begets flesh; spirit begets spirit. Wallis offers that Jesus might just have been half-alien.

Not only is Wallis happy to embrace a larger extended family, but Wallis also hopes that the aliens open up their technology goodie bags and hand us zero-point energy, a way to forego fossil fuels and nuclear energy. Intergalactic space travel sounds fun, he thinks.

Wallis falls back on the teaching of Jesus as God (or a group of ancient and highly evolved aliens?) infuses him with power as his son, bringing love, joy, power, and freedom for humans and their new brothers and sisters from the stars.

CONCLUSION TO ALIEN SECRETS

Francis Crick gave us a testable and logical theory for how life originated on Earth that boasts solid evidence. When your kids, or anyone, asks you how life started, you can answer "from the evolution of off-planet microbes based on synthetic DNA."

In review, Nobel prize-winning physicist Francis Crick presented a convincing argument, as co-discoverer of the structure of DNA, that DNA, the apparent source for all life on Earth, came not originally from Earth. Instead, he believes that DNA produces sophisticated genetic and mathematical bioengineering from an ancient civilization. Science suggests that aliens started life on Earth and that you and I are a result of alien synthetic DNA interacting with the Darwinian environment of Earth. We, and all Humans, are "Aliens from Earth. "First, it required hundreds of years of the development of science through the scientific method. Then it necessitated the twenty-seven-year completion of the Human genome project—no easy task. Then it took thirteen years of studying it by two exceptional scientists to provide the first evidence of mathematical coding that could be unlocked or activated to tell us our DNA comes from an intelligent design by aliens. An informal analysis of microbes and their evolution as performed in this book is consistent with life developing first off-planet. Nobel laureate Francis Crick proposed the exogenesis theory over three decades ago from the orderly and logical perfection of DNA structure, coding, and function but did not have the evidence, as we do now, to convince many of his peers. Studies must be continued for biological SETI.

After more than four billion years of life, we just now boast of technology and intellect capable of telling us our true origins.

We still do not know who or what is personally responsible. Still, if the SETI studies described in this book on the genetic code hold up and expand to confirm the message, life on Earth did not first evolve on Earth. Instead, it sprang from the genetic labs of one or more advanced ancient alien civilizations. Life was not born on Earth. A fictitious god did not create life. It spread here as bioengineered products to embed our environments and mold our planet into a most beautiful one teeming with life.

Our atoms come from earlier times in the Cosmos, some perhaps as old as thirteen billion years. Why not an ancient beginning for genetic coding, say up to eight billion years ago, or even ten? Maybe life did not need to face the tremendous challenge of evolving from scratch in primordial soups in most cases.

In addition to being Earthlings with an alien history, the above argument suggests that many beings across the galaxy may have arisen from synthetic DNA. We may have one huge galactic family of distant cousins with common origins. This is a testable theory. Just provide the genetic material from as many inhabited planets as you can and sit back and wait for confirmation.

Beginning over 6,000 years ago with the first civilizations such as Sumeria, clergyman Paul Wallis informs us in Escaping from Eden that myths involving ruling councils of aliens suggest that not only that aliens influenced history, including starting wars but reigned as the first perceived gods. Looking closely at Genesis and other biblical texts shows that Abrahamic and Hindi concepts of gods arose from alien myths. The Abrahamic religions' fabrication of a monotheistic god became a self-serving twist on tales of "Alien Councils."

There is at least one period in Human evolution in which the Great Apes, and only the Great Apes, benefited from abrupt increases in brain size that passed over the rest of the animal kingdom. Some suspect alien interventions.

Whether we want to admit it or not, we have been "aliens" on our planet all along. If Humans seeded life on Mars from our genetic designs, we would call it "Martian," even though it had "Human alien" elements. The same goes for us as "Earthlings" and the life on most other planets. The identity of our ancient makers may have been revealed in the detailed mathematical SETI analysis of our DNA by shCherbak and Makukov—and promptly censored and left unpublished.

In the same line of reasoning, we now possess the technology, as demonstrated so magnificently by the United States Navy, to track and record UAP in real-time. Not only did the released videos confirm the likely presence of aliens, but they also gave us the basics of how they must propel themselves and follow the laws of physics. This may lead to further insight into quantum mechanics and help us eventually develop similar technology with its spin-off applications.

Current "spin-off" applications for the aliens likely include beaming objects up or down, interfering with our consciousness, and disarming electronic devices, including missiles.

As the ultimate scientists, aliens demonstrated their cloning abilities with the little gray aliens retrieved from the Roswell crash. Abductees report incidents of impressive healing. There is probably little they cannot achieve biologically, and especially if supported by applied quantum mechanics.

We spoke at length about the myths and artwork records of aliens as ancient leaders, rulers, and perceived gods changing the course of history. These myths arose in Sumeria and were plagiarized and twisted to form part of Hinduism, Judaism, Islam, Christianity, and other religions.

We have good news and bad news about abductions of Humans by aliens. Let's start with the bad news. Abductions, against peoples' will by definition, seemed to flourish after the grays landed during WWII and, according to abductees, most often included operations and projects related to hybridization. When we consider the grays altogether, over fifty percent of abductees encountered grays. Colonel Corso claims we engaged some of the grays in a war but played down those statements in his Coast-to-Coast interviews.

The good news is that many of the other beings the abductees perceived seemed Human, as high as forty percent. Maybe twenty percent of those looked like Human-Alien hybrids.

We do not know yet who our ancient alien seeders were and whether they are represented in the aliens visiting us now—or if they even survived at all.

Still, in this book, we projected classic Darwinian behaviors that populations tend to exhibit to protect and spread their gene pools or "identity pools" if they have high AI affiliations. Alien game plans would likely follow what we observed throughout our history. Part of its about perceived identity and "tribes" and competition for scarce resources, territories, and proliferation for us through sexual intercourse. War, even though we all had the same origins, was the rule rather than the exception. Would the aliens be any different?

We talked about merging with technology and the advent of sophisticated AI through quantum computers. Worlds populated with genetically engineered humanoids and cloned cyborgs connected to extremely smart quantum AI maybe be we are all headed to compete successfully.

Our visitors are showing us a form of that already, our likely future.

We "primates" must merge either genetically or technologically, or both ways, or face extinction.

More alien secrets need to be revealed for our continued life on Earth.

Based on contact experiencers, the aliens have already begun hybridization, and likely long ago. Is the timing right for more to "ascend" into the alien worlds?

STAGES OF ASCENSION

We now exit from the scientific method mode to the scientific, logical, speculation mode, like we had to do in Section: Agendas and Identities. We guess, logically, based on what we have learned and what makes sense from Darwinian behavior and Earth history. It would seem logical for aliens to adopt humanoid forms for efficient influence in a world where people kill others based on their religion or appearance. From our genetic "programming," we may never accept the arrival of most aliens—unless they look very human. Darwinism taught us to treat anything that seemed radically different from our family as a threat. We naturally fear the "different." It worked and is hard to overcome.

STAGE ONE

We have already suggested what Stage One might look like. From the bones the aliens have thrown us, like crashed spaceships with occupants, and allowing their craft to be filmed and tracked in detail, we may finally figure out how quantum mechanics, gravity, and consciousness all come together. Applying that knowledge would lead to great, and perhaps inexpensive, advances in medicine, technology, energy production, and other fields. The "heal me quickly" machines we see in science fiction would become science fact. One quantum "AI" doctor could treat millions efficiently. Do you think our health care industry would like that to happen?

STAGE TWO

Unlike our state of being earlier over four hundred years ago, we have become a ripe target for mass hybridization. Our growing AI and weaponry, our advances in nanotech and genetics, may be cause for alarm for our alien friends. They may want to breed control mechanisms or "units" into our populations that become protectors of their empires. Earth is, of course, one of them.

Darwin would argue that spreading one's identity through genes and contributions to consciousness and societies for the more intelligent species is a strong urge that arose from natural selection as the ultimate survival mechanism.

Darwin would award it the gold medal if one species could spread itself to many planets and in many lifeforms. Whichever alien group seeded Earth with their DNA, and probably millions of others over four billion years ago, is way ahead of the game. To start, we should pick up the baton and do our share for the Moon and Mars, and space stations.

Human-looking aliens would share much of our genome to appear like us but would receive genetic enhancements for:

- Longevity
- Robust health and immune systems
- Expanded and accelerated intellect with connection to quantum AI
- Expanded physical strength
- Full control of their emotions and those around them
- Telepathy, including long-distance bordering on the nonlocal scales.

- Techno-implants to boost any of the above and communicate performance into quantum databases.
- They might be empowered with access to electromagnet field generators, the same technology that runs their spaceships and permits them to move heavy objects through space.

We can assume they would be educated to be devoted to the alien cause to help create an alien-based totalitarian state. They would likely infiltrate the Deep States and governments.

Yes, they would be a "super-race." Get over it. Whom do you call upon to save a planet from catastrophe, "Super Man." When you find someone who makes better, well-informed, logical decisions in the species' best interest at large, you put them in charge?

A super-race doesn't have to start killing off whoever is deemed inferior as in a Hitler Germany. If that were in the plan, most of us would already be gone. Instead, we passed the eight billion mark a while ago.

STAGE THREE

Likely earned from exemplary accomplishments as a Stage Two, we might expect Stage Three human hybrids to be enhanced with greater abilities to defend themselves because of their high value and to be wired into appropriate quantum intelligence computers like Stage Two. Rather than controlling an individual's mind or small group of people, they might control large groups.

Perhaps they would have the same repertoire of skills as Stage Two, just at a higher level, and also a leadership psyche. They would become the ultimate leaders on Earth, at least in public, and might lead the Star Trek types of the expedition.

SECTION: BACKMATTER

ET-BASED ANSWERS TO LIFE'S BIGGEST QUESTIONS

The following arose from what we covered in this book and opinions and speculation based on my other books. It's logical philosophy based on science.

(1) ARE WE ALONE?

No way. Aliens genetically engineered DNA and seeded Earth and probably the entire galaxy. There is robust evidence for Earth, and it makes sense that an alien civilization would seed more than one planet and likely many more. We see remnants of early alien viruses in the "extremely strange and ancient" Giant Viruses of today.

(2) WHAT IS THE ORIGIN OF LIFE?

For Earth, an alien lab. The microbes were either directed towards Earth or arrived on a natural carrier like a meteorite. DNA shows a mathematical logic to it and bears a mathematical code that researchers can unlock. It is a ten trillion to one probability that DNA arose from genetic engineering.

(3) WHAT IS LIFE'S MEANING?

It's personal for humans, but life tends to do anything to survive and propagate its identity, for example, its genomes. Our ultimate meaning is to survive and contribute beneficial ways to the universe as a habitat for life as a species. It's all about teamwork to keep life going for its own sake. Life forms a web of nonlocal consciousness.

(4) WHO OR WHAT IS "GOD?"

The collective rhythm and memory of the entire universe and all that came before it and will come after it. (Pantheistic view) Some aliens, and future humans, may demonstrate God-like powers. Something cannot be "God" without being the universe itself. All life forms are connected through quantum nonlocal effects that arise from DNA. Aliens seem to be able to amplify their telepathy through technology.

(5) IS THERE AN AFTERLIFE?

There is no scientific proof. Yet, but many will be possible through technology one day in virtual worlds and new physical forms.

(6) WHAT LIES AHEAD?

Our interaction with aliens, friends, and foes will shape at least part of us into forms similar to aliens so we can compete, or we will perish or become enslaved. We will further diversify our skillsets by genetic enhancement and physically merging with technology. Some of us will be engineered for space travel. One main survival tool is to spread one's identity as much as possible.

Colonizing the moon and Mars become logical first steps.

ABOUT SCOTT CAMPBELL

Investigative journalist Scott Campbell serves readers as the author of dozens of summary books, audiobooks, hundreds of blogs, and full-length books.

Best Seller Summaries offer an easy, painless, and affordable way to keep up with the world and its hottest non-fiction topics. Choose from ebook, printed, and audio formats.

For a comparable price and time investment, buy three summaries of bestsellers instead of one best-selling full-length book, and benefit from three times the breadth of knowledge.

Scott Campbell earned a Biology degree, Summa Cum Laude, a physiology degree, and a doctorate from prestigious universities. His signature skills center around his ability to recognize intricate patterns that underlie a variety of fields: science, technology, politics, behavioral psychology, ideologies, social studies, evolution, philosophy, financial markets, and government. These subjects drove his research starting in 2008.

Summaries1000

Printed in the United States of America

First Printing: October 2021

ENDNOTES AND REFERENCES

1 The God Equation> Michio Kaku

2 https://www.quantumconsciousness.org/content/overview-sh

3 Life Itself . Francis Crick.

4 The Day After Roswell . Philip Corso.

5 Escaping from Eden . Paul Wallis.

6 The Alien Agendas . Richard Dolan.

7 The Day After Roswell . Philip Corso.

8 The Day After Roswell . Philip Corso.

9 https://www.lewrockwell.com/2013/02/david-martin/stalins-secret-agents-the-subversion-of-roosevelts-government/

10 https://www.history.com/news/j-allen-hynek-ufos-project-blue-book

11 https://the-take.com/read/how-did-close-encounters-of-the-third-kind-receive-input-from-astronomer-dr-j-allen-hynek

12 https://www.trumanlibrary.gov/education/presidential-inquiries/establishment-cia

13 Statement by the Department of Defense on the Release of Historical Navy Videos" . U.S. Department of Defense . Retrieved 2020-04-28 .

14 https://www.nbcnews.com/news/us-news/leaked-navy-video-appears-show-u-f-o-california-coast-n1267688

15 Cooper, Helene; Kean, Leslie; Blumenthal, Ralph (2017-12-16). "2 Navy Airmen and an Object That 'Accelerated Like Nothing I've Ever Seen ' " . The New York Times Co . Retrieved 2020-05-14 .

16 UFOs regularly spotted in restricted U.S. airspace, report on the phenomena due next month" . CBS 60 minutes . Retrieved May 18, 2021

17 https://www.popularmechanics.com/military/research/a33413777/pentagon-ufo-program-materials-vehicles/

18 Henandez, Rey, Beyond UFOs: the Science of Consciousness and Contact with non-human Intelligence (Volume One), 2018.

19 https://www.exostudies.org/updates/33-types-of-non-human-intelligences-nhis/

20
file:///C:/Users/user/OneDrive/DOLAN/FREE%20STUDY/A_Study_on_Reported_Contact_with_Non-Human_Intelli.pdf

21 https://youtu.be/I4IJxjtawl0

22
https://www.sciencedaily.com/releases/2020/06/200616100831.htm

23 The Alien Agendas by Richard Dolan

24 Escaping From Eden by Paul Wallis

25 https://www.brainfacts.org/thinking-sensing-and-behaving/learning-and-memory/2013/changing-your-mind---scientists-disrupt-specific-memories-in-people

26 https://nautil.us/blog/are-the-brains-electromagnetic-fields-the-seat-of-consciousness

27
https://www.tandfonline.com/doi/full/10.1080/17588928.2020.1839037

28
https://www.quantumconsciousness.org/content/overview-sh

29 Roger Penrose, Stuart Hamerof. Orch OR theory

30

The Alien Agendas . Richard Dolan.

31 Cognitive Neuroscience Current Debates, Research & Reports

Volume 12, 2021 - Issue 2

32 https://www.resonancescience.org/blog/A-New-Study-Examines-How-Consciousness-in-the-Universe-is-Scale-Invariant-and-Implies-an-Event-Horizon-of-the-Human-Brain?fbclid=IwAR2w5YRngKviua0P6AlvhUpu_RpmCGjjHwuGRBILbT3_LeXAL72V6VsXWxk

33 Emerson, D. J., Weiser, B. P., Psonis, J., Liao, Z., Taratula, O., Fiamengo, A., Wang, X., Sugasawa, K., Smith, A. B., Eckenhoff, R. G., & Dmochowski, I. J. (2013). Direct modulation of microtubule stability contributes to anthracene general anesthesia. Journal of the American Chemical Society, 135(14), 5389–5398. Crossref PubMed.

34 Craddock, T. J. A., Kurian, P., Preto, J., Sahu, K., Hameroff, S. R., Klobukowski, M., & Tuszynski, J. A. (2017). Anesthetic alterations of collective terahertz oscillations in tubulin correlate with clinical potency. Scientific Reports, 7(1), 9877. Crossref PubMed.

35 https://www.psychologytoday.com/us/blog/hypnosis-the-power-trance/201406/hypnotic-regression-and-healing-the-unconscious-mind

36 https://web.archive.org/web/20060518075852/http://homepages.ihug.co.nz/~sai/pribram.htm

37 https://www.newscientist.com/article/dn23343-interspecies-telepathy-human-thoughts-make-rat-move/

38 https://youtu.be/ZuVsVeYK_iw

39 https://www.researchgate.net/publication/237454651_Videotaped_experiments_on_telephone_telepathy

40 https://www.cnbc.com/2021/04/09/elon-musks-neuralink-shows-video-of-monkey-using-mind-to-play-pong.html

41 https://fountainmagazine.com/2011/issue-84-november-december-2011/quantum-worlds-from-entanglement-to-telepathy

42 http://cogprints.org/3065/1/qmt.pdf

43 Alien Agendas by Richard Dolan

44 https://www.euroweeklynews.com/2021/10/11/ex-us-military-chief-claims-aliens-tampered-with-nuclear-missiles/

45 https://www.lehigh.edu/~dwp0/Assets/images/Is%20the%20answer%20to%20life,%20the%20universe%20and%20everything%2037%3F%20-%20space%20-%2022%20December%202014%20-%20New%20Scientist.pdf

46 https://www.sciencedaily.com/releases/2010/08/100817122405.htm

47 https://en.wikipedia.org/wiki/DNA_teleportation

48 https://www.techjuice.pk/breakthrough-as-scientists-successfully-teleport-an-object-to-space/

https://www.techjuice.pk/breakthrough-as-scientists-successfully-teleport-an-object-to-space/

49 https://en.wikipedia.org/wiki/DNA_teleportation

50 https://pubmed.ncbi.nlm.nih.gov/26098521/

51 https://www.npr.org/sections/krulwich/2010/05/28/126510251/aliens-found-in-ohio-the-wow-signal

52 https://newstarget.com/2020-11-19-spacefaring-bacteria-survived-full-year-in-space-says-researchers.html

53 (Elliott, 2010)

54 https://pubmed.ncbi.nlm.nih.gov/12941485/

55 Could Giant Viruses Be the Origin of Life on Earth?

The ancestors of modern viruses may have laid the groundwork for cellular life as we know it. Quanta Magazine. Carrie Arnold. July 17, 2014.

56 . Koonin, Eugene. June of 2014. Microbiology and Molecular Biology Reviews.

57 https://www.sciencedirect.com/science/article/abs/pii/S0923250805002676# !

58 https://www.nature.com/articles/s41467-020-15507-2

59 https://www.sciencedirect.com/topics/immunology-and-microbiology/giant-virus

60 http://www.frangun.org/forterre_MedecineSciences_2005.pdf

61 http://bip.cnrs-mrs.fr/bip10/LUCA.pdf

62 Life before LUCA * Athel Cornish-Bowden and María Luz Cárdenas

63 https://news.illinois.edu/view/6367/204989

64 https://bmcecolevol.biomedcentral.com/articles/10.1186/1471-2148-12-156

65 https://australian.museum/learn/science/human-evolution/homo-sapiens-modern-humans/

66 https://www.cambridge.org/core/journals/international-journal-of-astrobiology/article/abs/seti-in-vivo-testing-the-wearethem-hypothesis/43E3302CCE1D053886F35C819CD5E55D

67 https://link.springer.com/chapter/10.1007/978-1-4020-6340-4_7

68 https://thebiomedicalscientist.net/science/synthetic-dna-extraterrestrial-intelligence-future-biomolecules

69 Brain Bugs: How the Brain's Flaws Shape our Lives. Dean Buonomano. August 2012.

70 https://www.pewresearch.org/fact-tank/2021/06/30/most-americans-believe-in-intelligent-life-beyond-earth-few-see-ufos-as-a-major-national-security-threat/

71 https://www.simonandschuster.com/authors/Philip-Corso/1246221

72 https://www.simonandschuster.com/authors/Philip-Corso/1246221

73 https://www.youtube.com/watch?v=BFdlwXw9tX0

74 https://en.wikipedia.org/wiki/Arthur_Trudeau

75 https://youtu.be/9siSobC6Gw0

76 https://youtu.be/a7I4Ag8gq70

77 The Day After Roswell by Philip Corso

78 https://youtu.be/7lVM9IdAdo0

79 Carey. Ufo Secrets Inside Write Patterson. 2019

80 https://youtu.be/uD2186Yh0Uc

81 https://www.smithsonianmag.com/smart-news/reagan-and-gorbachev-agreed-pause-cold-war-case-alien-invasion-180957402/

82 https://www.youtube.com/watch?v=VgziDyPSUog

83 https://newstarget.com/2020-10-11-space-force-agreement-military-moon-base.html

84 https://spacenews.com/china-russia-reveal-roadmap-for-international-moon-base/

85 https://www.latimes.com/archives/la-xpm-1989-07-18-mn-4091-story.html

86 https://youtu.be/DFPUq_nE_Hs

87 https://link.springer.com/chapter/10.1007/978-1-4020-6340-4_7

88 https://www.wired.com/2006/06/ufo-hacker-tells-what-he-found/

89 https://www.sciencedaily.com/releases/2018/04/180411131659.htm

---

[i] https://knowledge-time.com/us-army-high-ranked-confirms-57-alien-species-that-look-like-us-and-walk-among-us/?fbclid=IwAR1chAh81r9mTTQV0Pchlq7QGT8N1tIeZilBCDya7QaQSEOhnB_uWNmdAi4

CPSIA information can be obtained
at www.ICGtesting.com
Printed in the USA
LVHW040726201222
735517LV00008B/520

9 781954 241268